Evolving Methods for Macromolecular Crystallography

NATO Science Series

A Series presenting the results of scientific meetings supported under the NATO Science Programme.

The Series is published by IOS Press, Amsterdam, and Springer in conjunction with the NATO Public Diplomacy Division

Sub-Series

I.	Life and Behavioural sciences	IOS Press
II.	Mathematics, Physics and Chemistry	Springer
III.	Computer and Systems Science	IOS Press
IV.	Earth and Environmental Sciences	Springer

The NATO Science Series continues the series of books published formerly as the NATO ASI Series.

The NATO Science Programme offers support for collaboration in civil science betweem scientists of countries of the Euro-Atlantic Partnership Council. The types of Scientific meeting generally supported are "Advanced Study Institutes" and "Advanced Research Workshops", and the NATO Science series collects together the results of these meetings. The meetings are co-organized bij scientitsts from NATO countries and scientists from NATO's Partner countries – countries of the CIS and Central and Eastern Europe.

Advanced Study Institutes are high-level tutorial courses offering in-depth study of latest advances in a field.
Advanced Research workshops are expert meetiings aimed at critical assessment of a field, and identification of directions for future action.

As a consequence of the restructing of the NATO Science Programme in 1999, the NATO Science Series was re-organised to the four sub-series noted above. Please consult the following web sites for information on previous volumes published in the Series.

http://www.nato.int/science
http://www.springer.com
http://www.iospress.nl

Series II: Mathematics, Physics and Chemistry – Vol. 245

Evolving Methods for Macromolecular Crystallography

The Structural Path to the Understanding of the Mechanism of Action of CBRN Agents

Edited by

Randy J. Read

Department of Haematology, University of Cambridge, Cambridge Institute for Medical Research, Cambridge, U.K.

Joel L. Sussman

Department of Structural Biology, Weizmann Institute of Science, Rehovot, Israel

 Springer

Proceedings of the NATO Advanced Study Institute on
Evolving Methods for Macromolecular Gystallography:
The Structural Path to the Understanding of the Mechanism
of Action of CBRN agents
Erice, Italy
19—28 May 2005

A C.I.P. Catalogue record for this book is available from the Library of Congress.

ISBN 978-1-4020-6315-2 (PB)
ISBN 978-1-4020-6314-5 (HB)
ISBN 978-1-4020-6316-9 (e-book)

Published by Springer,
P.O. Box 17, 3300 AA Dordrecht, The Netherlands.

www.springer.com

Printed on acid-free paper

TABLE OF CONTENTS

PREFACE

This volume comprises papers presented at the 2005 edition of the "Crystallography of Molecular Biology" courses that have been held since 1976 at the Ettore Majorana Centre for Scientific Culture in Erice, Italy. This series of courses is renowned for bringing leaders in the field of macromolecular crystallography together with highly motivated students, in a beautiful and intimate location that encourages people to interact. The warm and informal atmosphere at these Erice conferences, especially these on crystallography, has helped to foster long-term scientific interactions and new international friendships that have often lasted for the lifetime of the scientists. The course was financed by NATO as an Advanced Study Institute and by the European Commission as a EuroSummerSchool.

The papers span the breadth of material presented in the course, which emphasized the practical aspects of modern macromolecular crystallography and its applications. One must start with crystals: Bergfors showed how to improve initial crystals through seeding, while Byrne discussed the difficult problem of crystallizing membrane proteins. The collection of optimal diffraction data requires both careful preparation of cryo-cooled crystals (Garman) and proper processing of the diffraction images (Leslie). To obtain images of electron density, one needs estimates of the phases of the diffracted spots. Sheldrick presented the background to the single-wavelength anomalous diffraction (SAD) method, which has been gaining popularity, and McCoy discussed the basis of modern maximum likelihood methods for treating information in experimental phasing. When a related structure is known, the phases can be obtained by molecular replacement, which can use stochastic search methods (Glykos) or tree search methods based on maximum likelihood (Read). There is also the promise that *ab initio* phasing methods will contribute at least at low resolution (Lunin). Initial phases can be improved dramatically by density modification (Turk). Increasingly, all these methods can be automated (Adams), an important step to increasing the throughput of structural genomics efforts (Baker). At times, structural genomics provides structures without a known function, but Thornton showed that structure alone can shed light on function. Careful analysis of structures can provide an explanation for disease processes at the atomic level (Jaskolski). The climax of this volume, as of the course, is the demonstration by Sayre that diffraction can be used to image single particles as large as cells.

Most of the real organizational work for the course was done by Paola Spadon and Lodovico Riva di Sanseverino, who, between them, found most of the funding, corresponded with applicants and selected participants, made the logistical arrangements, and reminded us patiently when we needed to do something. Lodovico brought a wealth of experience to bear, having been a mainstay of the Erice meetings since their inception. John Irwin played an essential role, organizing all the information technology (IT) facilities needed to conduct tutorials and demonstrations in a computer (CPU)-intensive field like macromolecular crystallography.

Paola, Lodovico, and John were joined as Fellows of the Loyal Order of Orange Scarves by a set of enthusiastic volunteers: Vito Calderone, Laura Cendron, Sonia Covaceuszach, Federica Morandi, Elena Papinutto, Nicola Pasquato, Fabiana Renzi, and Donatella Tondi. Together they dealt with any of the day-to-day emergencies that arise in running a course like this.

In addition to the essential support from NATO and the European Commission, generous financial support was received from the International Union of Biochemistry and Molecular Biology, INTAS, the International Union of Crystallography, the University of Bologna, AstraZeneca, CCP4, and Douglas Instruments.

Randy J. Read and Joel L. Sussman

SUCCEEDING WITH SEEDING: SOME PRACTICAL ADVICE

TERESE BERGFORS

Department of Cell and Molecular Biology, Uppsala University
Biomedical Center, Box 596, 751 24 Uppsala, Sweden

Abstract: Seeding is a powerful and versatile method for optimizing crystal growth conditions. This article discusses, from a practical point of view, what seeding is, the selection and transfer of seeds, and into what conditions they should be transferred. The most common causes of failures in seeding experiments are also analyzed.

Keywords: crystallization; microseeding; optimization; seeding; streak seeding.

1. Introduction

Crystallization is the rate-limiting step in the process of determining a three-dimensional macromolecular structure by X-ray crystallography. Automation and miniaturization of the crystallization setup have greatly facilitated massive screening. However, screening in itself, no matter how extensive, is still no guarantee that crystals will be found or that they will provide diffraction-quality data. At least half of the crystals obtained in an initial screen cannot be used without further optimization [1]. Therefore, optimization methods are often crucial for the success of a crystallization project.

One powerful tool in the arsenal of optimization techniques is seeding. While it is not a universal solution to all optimization problems, seeding is relatively cheap, fast, and easy, which makes it worth trying at an early stage. Possible applications include:

- If spontaneous nucleation is slow, i.e., the drop stays clear for a long time (weeks to months) before crystals appear

- To reduce showers of crystals

- To increase the size of crystals

- To improve reproducibility due to erratic nucleation, i.e., supposedly identical drops do not consistently produce crystals

- If crystals grow in clusters rather than singly

1

R.J. Read and J.L. Sussman (eds.): Evolving Methods for Macromolecular Crystallography, 1–10
© 2007. *Springer*

- To improve crystal quality
- As a diagnostic tool to determine if the drop is undersaturated or supersaturated.

2.   What is seeding?

Seeding is the use of an ordered, solid phase which is introduced into an experiment to act as a growth surface for the crystals. Spontaneous nucleation (the generation of a stable, ordered nucleus) is a kinetically demanding step. Therefore, molecules prefer to accumulate on a ready-made template, if one is available.

2.1. HOMOGENEOUS VS HETEROGENEOUS NUCLEANTS

If the seed consists of the same molecules as the target to be seeded, it is said to be a homogeneous nucleant. The molecules need not be identical though crystals of one protein can be used as seeds for the crystallization of a related form of the protein, e.g., a homologue from a different species, the selenomethionyl-substituted form, a mutated or slightly different construct, or in complex with a ligand, heavy atom, or cofactor. Generally speaking, the more similar two proteins are, the more likely it is that crystals of one will be able to function as a template for the crystal growth of the other. (Were this not the case, lysozyme crystals would work as seeds for recalcitrant proteins that refuse to crystallize. Alas, things are never that easy.)

Nevertheless, heterogeneous nucleants, i.e., materials consisting of molecules unrelated to the target protein, can be extremely effective. Anyone who has encountered crystals growing on a clothing fiber in their drops [2] or on a scratch in the glass coverslip has already experienced heterogeneous nucleation. The inclusion of these nucleants is usually unintentional. A universal substance, capable of catalyzing the crystallization of all proteins, is the dream of every crystallographer, but like the philosopher's stone in alchemy, its existence remains elusive. Some of the many materials that have been tested as heterogeneous nucleants in protein crystallization include hair [2], sand [3, 4], lipid layers [5], polyvinylidene difluoride membranes [6], porous silicon wafers [7, 8], and Langmuir-Schaeffer films [9].

This article will focus chiefly on microseeding techniques using crystals of structurally related or identical proteins as the seeds.

3.   What to use as seeds

Any of the following can be used as nucleants or seeds:

- Single, small crystal, typically 0.05–10 microns in the longest dimension.
- Slurry of microcrystals.

- Fragment chipped from a larger crystal.

- Crystal, or pieces thereof, that has already been used in the X-ray beam.

- Crystalline precipitate.

- In desperate cases, any solid phase of the protein, e.g., gels or oils [4, 10]. Gels and oils are not ideal starting material, but they do exhibit some short-range order and this may be enough to trigger the ordered growth of a crystal.

Seed quality is dependent on the quality of the parent crystals. Large crystals do not make good seeds because they have accumulated too many defects. It is therefore better to chip a fragment from the large crystal or to smash it into small fragments; this will generate fresh edges and surfaces for growth. Extremely small crystals and crystalline precipitates can be used without pulverization; crystals typically referred to as sea urchins, plates, etc. can be smashed or crushed in the drops where they have grown and used directly from there.

3.1. PREPARING AND STORING SEED STOCK SOLUTIONS

It is often more convenient to work with the seeds in a microcentrifuge tube instead of working directly with them from the droplet. Among other things, the droplets tend to dry out or one may wish to make a quantitative dilution series (see Section 4) of the seeds and reuse them. Therefore, to make a seed stock, the parent crystals are transferred in some of their mother liquor to a microcentrifuge tube. Glass tissue homogenizers, Seed Beads (Hampton Research, Inc.), sonication, vortexing, acupuncture needles, dentist's tools, etc. can be used to crush the crystals into a crystal slurry [11]; the actual method of generating the seeds is not critical. What is important is that the seeds are collected and stored in a mother liquor where they are stable, i.e., do not dissolve or become contaminated by microbial growth. A common mistake is that seeds that have been grown at room temperature are transferred to a microcentrifuge tube and put in the refrigerator. Not all crystals tolerate moving back and forth between the refrigerator and the lab bench. Some people flash-cool their seed stock in liquid nitrogen [12]; here again, not all seeds may survive this treatment. Thus, for the storage and reuse of seed stocks, the following simple precautions should be taken:

- Remove a 5–10 µL aliquot of the seed stock and examine it under the microscope at high magnification to verify that the seeds have indeed survived the storage conditions. This is especially important to do if temperature changes have been involved.

- Check the seed stock solution for the odor of bacterial contamination and if necessary, make new mother liquor and seeds.

4. How to transfer the seeds

The use of an intact, single crystal as a seed is referred to as macroseeding. This is a difficult technique. (For those who wish to attempt it anyway, an excellent how-to example is found in Mowbray [13].) Working with microseeds is much easier, especially by streak seeding [10, 14–16]. Streak seeding usually employs some kind of animal hair or whisker as the transfer tool (a so-called seeding wand) for the seeds. The surface of an intact parent crystal in a droplet is stroked with the hair to pick up microcrystals. Alternatively, the seeding wand can be dipped into a droplet or microcentrifuge tube containing a slurry of microcrystals. In either case, the microcrystals or seeds adhere to the hair, which is then swiped through the new crystallization drop whereby the seeds are transferred. The new crystals will grow along the streak line of deposited seeds (Figures 1a and b).

This is by far the simplest and fastest way of seed transfer, and the crystals that result may well be large enough to use without further refinement of the seeding protocol.

However, it may be necessary to reduce the number of seeds; this can be done, for example, by passing the seeding wand through several washes of mother liquor to remove the excess seeds. A more quantitative method is to make a dilution series of the seed stock. Two detailed descriptions of how to make seed dilutions can be found in Fitzgerald and Madsen [17] and Luft and DeTitta [11].

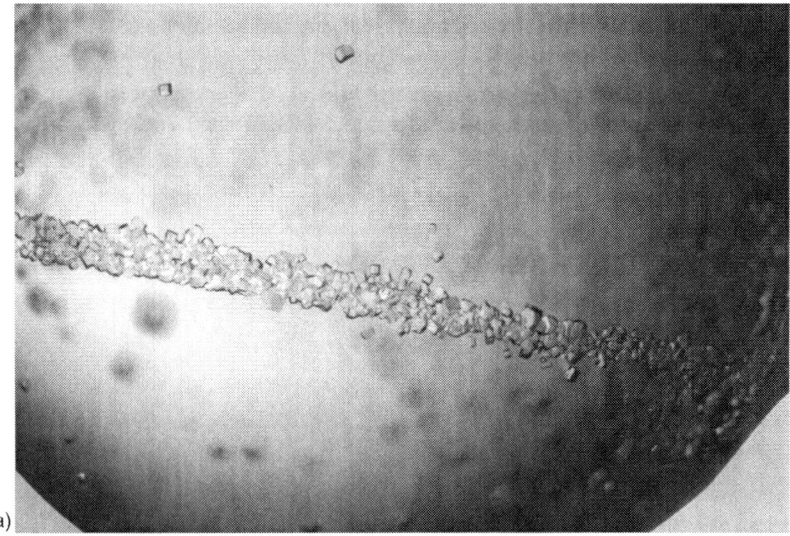

(a)

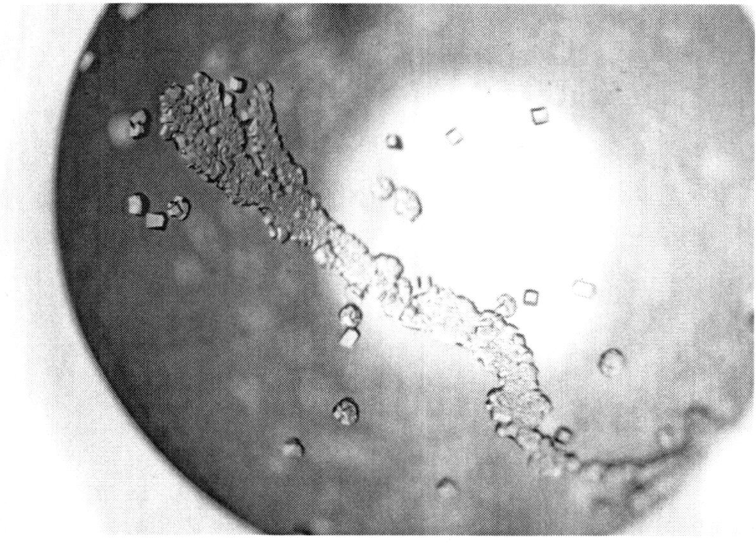

(b)

Figure 1. Lysozyme crystals growing on streak seeding lines: (a) vapor-diffusion setup; (b) microbatch-under-oil. The oil does not affect the seed transfer.

5. What to seed into: supersaturation levels in the new drop

Some understanding of seeding in terms of the phase diagram is helpful when deciding on the composition of the new drops where the seeds are to be deposited.

For a solution to crystallize, it must first be in a state of supersaturation. This in itself is no guarantee that crystals will form: crystallization also requires nucleation. This event is the formation of the first stable, ordered nucleus. The likelihood of this happening is related to the number of molecules in solution – the more there are, the greater the probability that any two or more of them will collide with each other, overcome the competing forces to drive them apart, and remain as an ordered nucleus.

Unfortunately, the levels of supersaturation that promote spontaneous nucleation are too high for the slow, accumulative growth that leads to well-ordered, large-sized crystals. For this reason, if nucleation does manage to occur, it often results in showers of small crystals rather than a few, single, and large ones. Seeding is an optimization technique that separates the nucleation event from the growth process: the seed crystals are removed from the original drop in which they nucleated and placed in a new experimental condition. The new drop should be equilibrated at a level of supersaturation high enough to support crystal growth, but low enough to prevent spontaneous nucleation. These different regions of supersaturation are represented in the phase diagram as the labile and metastable zones (Figure 2a).

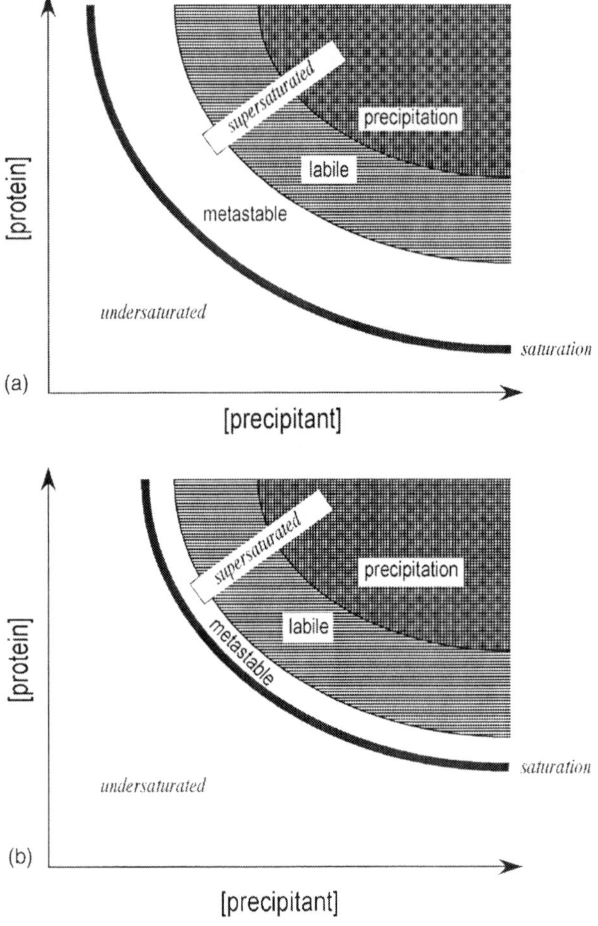

Figure 2. (a) A theoretical phase diagram. The thick black line represents the limit of solubility. Below this line, the solution is undersaturated and above it, supersaturated. Seeds placed in an undersaturated solution dissolve. Spontaneous nucleation occurs in the labile zone of the supersaturated phase, but the best crystal growth occurs in the metastable region. At the highest levels of supersaturation, the solid phase aggregates in a disordered fashion, seen as precipitate. (b) The same as part (a), but here the metastable region is very narrow, making it a poor candidate for seeding experiments.

In practice, the phase diagram of the protein is rarely available. The location of the different zones within the supersaturated phase has to be determined empirically unless the crystallization setup is equipped with some kind of instrumentation, e.g., a light-scattering instrument, to monitor the onset of nucleation [18]. When the problems are too much or too rapid nucleation, the concentration of the protein or the precipitant, or both, should be lowered in the new set of drops to be seeded. The following rule of thumb can be used as a starting point in the search for the metastable zone: begin by halving the protein concentration in the new drops. Of course, precipitant concentration could also be manipulated, but lowering the protein concentration has a slight advantage in that it reduces the amount of protein sample needed. The ease with which the metastable zone can be localized will depend on how wide it is (Figure 2b). Some proteins have very narrow metastable zones. In these cases, it is almost impossible to pinpoint the right concentrations of protein or precipitant for adding the seeds and some other optimization method will be necessary.

When the problem is slow nucleation, i.e., the drops stay clear for weeks or months before the crystals appear, adding a seed will jump start the nucleation process. The same or even higher protein or precipitant concentrations can be used in the new drops because the goal is not to reduce nucleation but to encourage it.

6. Troubleshooting

When seeding fails to work as expected, it is often due to an improperly designed experiment or mishandling of the seeds. Some of the common problems that can sabotage the success of a seeding experiment are described below.

6.1. THE EXPERIMENT IS NOT EQUILIBRATED

Adding seeds to an undersaturated solution will cause them to dissolve. Many vapor-diffusion experiments begin in an undersaturated state and only reach supersaturation after a period of equilibration against their reservoirs. It is difficult to give a general rule for how long time one should wait before seeding the new drops: the time required for the drop to equilibrate with the reservoir is dependent upon many different factors (see Luft and DeTitta [19] for a review and for specific examples [20].) However, as a rough guide, the effects of seeding should be obvious within 2–7 days. If the drops that have been seeded are still clear after 1 week, the cause may well be that the seeds dissolved.

6.2. EVAPORATION HAS OCCURRED

The drop may have been properly equilibrated to, or have been begun at, the metastable zone, but when the experiment is reopened for the purpose of adding the seeds, some evaporation occurs and the level of supersaturation soars back into the labile zone. This is always a potential problem in vapor-diffusion experiments and is more pronounced with hanging drops than sitting drops, and with extremely small drop volumes. In this respect, seeding into microbatch is more reproducible.

6.3. NO SEEDS ARE TRANSFERRED

Improper storage of the seeds, as discussed in Section 3.1, may cause the seeds to dissolve. Another scenario is that all the seeds have sunk to the bottom of the microcentrifuge tube. The seed stock should be vortexed immediately before use to redisperse the seeds. In transfers by streak seeding, the seeds may not have adhered to the seeding wand. The animal whiskers or hairs do wear out as seeding wands and need to be replaced when they no longer seem to be depositing seeds. Acupuncture needles are another popular tool for seed transfer, but it should be kept in mind that smooth surfaces like metal and glass do not trap the seeds nearly as well as hair.

6.4. TOO MANY HETEROGENEOUS NUCLEANTS ARE PRESENT

Too many heterogeneous nucleations already present in the drop will mask the effect of any intentionally added nucleant, i.e., the seeds. Typical extraneous sources of nucleation are dirt, dust, denatured protein molecules, and clothing fibers. If seeding is not giving the desired effect, filter all the components of the drops through a 0.22 μm filter immediately before setup. Wear a lab coat and work cleanly.

6.5. THE PROTEIN IS NOT PURE ENOUGH

Seeding is not particularly effective in improving crystal quality if the problem is microheterogeneity in the protein sample [21]. An additional purification step may be required instead of, or prior to, application of the seeding protocol. Microheterogeneity can be assayed, for example, with isoelectric focusing gels.

7. Summary

Although efforts to automate crystal optimization are now underway [1], a follow-up optimization experiment is often still designed and implemented manually. Microseeding, and especially streak seeding, are easy to perform which makes them attractive as optimization methods. Seeding has a wide range of applications and the recommendations presented here will hopefully encourage its implementation in the laboratory.

Acknowledgments and disclaimer

I thank my colleagues in Uppsala and Stockholm for their input on the manuscript. The inclusion of this paper in the NATO Science Programme series is not to be construed as an endorsement by this author of NATO, nor may any of the material in this article be used for a military purpose.

References

1. Chayen, N. and Saridakis, E. (2002) Protein crystallization for genomics: towards high-throughput optimization techniques. *Acta Crystallographica*, **D58**: 921–927.
2. D'Arcy, A., MacSweeney, A., and Haber, A. (2003) Using natural seeding material to generate nucleation in protein crystallization experiments. *Acta Crystallographica*, **D59**: 1343–1346.
3. Hendrickson, W., Horton, J., and LeMaster, D. (1990) Selenomethionyl proteins produced for analysis by multiwavelength anomalous diffraction (MAD): a vehicle for direct determination of three-dimensional structure. *The EMBO Journal*, **9**: 1665–1672.
4. Bergfors, T. (2003) Seeds to crystals. *Journal of Structural Biology*, **142**: 66–76.
5. Hemming, S., Bochkarev, A., Darst, S., Kornberg, R., Ala, P., Yang, D., and Edwards, A. (1995) The mechanism of protein crystal growth from lipid layers. *Journal of Molecular Biology*, **246**: 308–316.
6. Punzi, J., Luft, J., and Cody, V. (1991) Protein crystal growth in the presence of poly(vinylidene difluoride) membrane. *Journal of Applied Crystallography*, **24**: 406–408.
7. Chayen, N., Saridakis, E., El-Bahar, R., and Nemirovsky, Y. (2001) Porous silicon: an effective nucleation-inducing material for protein crystallization. *Journal of Molecular Biology*, **312**: 591–595.
8. Sanjoh, A., Tsukihara, T., and Gorti, S. (2001) Surface-potential controlled Si-microarray devices for heterogeneous protein crystallization screening. *Journal of Crystal Growth*, **232**: 618–628.
9. Pechkova, E. and Nicolini, C. (2002) Protein nucleation and crystallization by homologous protein thin film template. *Journal of Cellular Biochemistry*, **85**: 243–251.
10. Stura, E. and Wilson, I. (1990) Analytical and production seeding techniques. *Methods: A Companion to Methods in Enzymology*, **1**: 38–49.
11. Luft, J. and DeTitta, G. (1999) A method to produce microseed stock for use in the crystallization of biological macromolecules. *Acta Crystallographica*, **D55**: 988–993.
12. Scheidig, A., Sanchez-Lorente, A., and Lautwein, A. (1994) Crystallographic studies on p21$^{H\text{-}ras}$ using the synchrotron Laue method: improvement of crystal quality and monitoring of the GTPase reaction at different time points. *Acta Crystallographica*, **D50**: 512–520.
13. Mowbray, S. (1999) Macroseeding: a real-life success story. In *Protein Crystallization*. Edited by Bergfors, T. La Jolla: International University Line, pp. 157–162.
14. Stura, E. and Wilson, I. (1991) Applications of the streak seeding technique in protein crystallization. *Journal of Crystal Growth*, **110**: 270–282.
15. Stura, E. and Wilson, I. (1992) Seeding techniques. In *Crystallization of Nucleic Acids and Proteins*. Edited by Ducruix, A. and Giege, R. Oxford: IRL Press, pp. 99–126.
16. Stura, E. (1999) Seeding. In *Protein Crystallization*. Edited by Bergfors, T. La Jolla: International University Line, pp. 141–153.
17. Fitzgerald, P. and Madsen, N. (1986) Improvement of limit of diffraction and useful X-ray lifetime of crystals of glycogen debranching enzyme. *Journal of Crystal Growth*, **76**: 600–606.
18. Saridakis, E. (2000) Optimization of the critical nuclear size for protein crystallization. *Acta Crystallographica*, **D56**: 106–108.
19. Luft, J. and DeTitta, G. (1997) Kinetic aspects of macromolecular crystallization. *Methods in Enzymology*, **276**: 110–131.

20. Luft, J. and DeTitta, G. (1995) Chaperone salts, polyethylene glycol and rates of equilibration in vapor-diffusion crystallization. *Acta Crystallographica*, **D51**: 780–785.
21. Caylor, C., Dobrianov, I., Lemay, S., Kimmer, C., Kriminski, S., Finkelstein, K., Zipfel, W., Webb, W., Thomas, B., Chernov, A., and Thorne, R. (1999) Macromolecular impurities and disorder in protein crystals. *Proteins: Structure, Function and Genetics*, **36**: 270–281.

EXPRESSION, PURIFICATION, AND CRYSTALLISATION OF MEMBRANE PROTEINS

BERNADETTE BYRNE

Division of Molecular Biosciences, Imperial College London, London, SW7 2AZ, UK

Abstract: Approximately, 29,000 protein structures are deposited in the Protein Databank (January 2005), but only about 90 of which are independent membrane protein structures. This represents a significant increase in knowledge compared with a matter of only 5 years ago when a mere handful of membrane protein structures were available. Despite the advances, our understanding of the structure–function relationships and mechanism of action of many membrane proteins is still lacking. This is particularly true of many of the more clinically relevant membrane proteins, such as the G-protein-coupled receptors (GPCRs). The GPCRs regulate cellular responses to a wide range of biologically active molecules including hormones and drugs and are thus important targets for therapeutic intervention in a number of disease states. However, the increasing number of membrane protein structures has provided a critical mass of information which has yielded a more rational approach to the process of obtaining diffraction quality crystals. It is the different stages of this process; expression, solubilisation, purification, and crystallisation that will be covered in this lecture.

Keywords: membrane protein; structure determination; expression; solubilisation; purification; crystallisation.

1. Expression of membrane proteins

Some membrane proteins are produced at very high levels in the cell including proteins involved in energy generation such as the respiratory and photosynthetic complexes and proteins such as rhodopsin, a G-protein-coupled receptor (GPCR). It is no coincidence that a large number of these have been extensively characterised both biochemically and structurally. However, most membrane proteins are produced at endogenously low levels (often as low as a few molecules per cell) and thus for any kind of structural analysis there is

11

R.J. Read and J.L. Sussman (eds.): Evolving Methods for Macromolecular Crystallography, 11–23
© 2007. *Springer*

the requirement that these proteins are expressed using recombinant systems. The use of such recombinant systems allows

- Controlled production of the target protein (theoretically). In principle it is possible to produce the maximum amount of target protein by varying the strength of the promoter and the time and temperature at which expression is initiated.

- Addition of an affinity tag to facilitate detection and purification of the protein of interest.

There are a wide range of expression systems from bacterial to mammalian available for the production of recombinant membrane proteins, each with associated advantages and disadvantages depending on the target protein. Relatively large quantities of the target protein are required, often on a tight budget, which can dictate both the choice of target protein and the system used. *Escherichia coli*-based expression systems are still the systems of choice due to their flexibility, ease of use, and cheapness. However, their use has proved limited, in particular, with respect to the most interesting eukaryotic targets. Other bacterial-based systems are also available which have had reasonable success in the production of membrane proteins, most notably *Lactococcus lactis* (see review by Kunji et al. [1]). Recently, a spectacular success has been achieved using the eukaryotic *Pichia pastoris*-based expression system for the production of the voltage-dependent Shaker family K + channel from rat brain [2].

Another important feature of recombinant expression systems is the ability to tag the protein of interest. It is possible to use both detection tags to allow monitoring of expression and affinity tags to facilitate purification. Combinations of tags are also possible. Table 1 gives a list of commonly used epitope tags. By far, the most common tag is the His tag comprised of a number

TABLE 1. List of epitope tags used for detection and purification of recombinantly expressed proteins

Tag	Suitable for detection?	Suitable for purification?
His	Yes	Yes
Haemagglutinin	Yes	No
Strep	Yes	Yes (only *Escherichia coli*)
Strep II	Yes	Yes
Biotin	Yes	Yes
Flag	Yes	No
Protein C	Yes	Yes
C-myc	Yes	No
V5	Yes	No
GST	Yes	Yes

of histidine residues in series. These can vary in number (6–10 usually) and position within the gene of interest (usually N- or C-terminus).

Despite a number of advances, the specific recombinant expression of membrane proteins still represents a major challenge. Many membrane proteins are cytotoxic to the cell when produced in high levels. Others form inclusion bodies, insoluble aggregates of the protein of interest. In most cases, this is a major disadvantage; however, inclusion bodies can be exploited as a source of concentrated membrane protein. It has been possible to refold and subsequently crystallise some β-barrel outer membrane proteins (OMPs) produced as inclusion bodies [3, 4]. It has also been possible to refold a number of α-helical proteins including bacteriorhodopsin and the light harvesting complex; however, this has yet to result in well-diffracting crystals [5]. Further work is needed in this area if we are to fully understand the processes involved in refolding and develop widely applicable refolding systems.

Another major frustration results from cleavage of the affinity tag from the protein of interest during expression. Detection of the protein is possible by Coomassie Blue staining of a standard sodium dodecyl sulfate polyacrylamide gel electrophoresis (SDS-PAGE) gel providing the protein is expressed at high levels. However, it can result in a protein which is difficult, if not impossible, to purify. This problem can be remedied by

- Moving the tag
- Using an alternative tag

However, neither solution is guaranteed to work.

One tried and tested method to the successful expression and ultimately crystallisation of membrane proteins relies on high-throughput approaches. A very nice example of this is the SecY complex whose structure was solved just over 1 year ago [6]. In this case, the researchers started with SecY homologues from ten different bacteria. The homologue from *Methanococcus jannaschii* was the one that ultimately yielded the high-resolution structure. This approach can make it possible to obtain structural information for one of a group of membrane proteins, but does not solve the issues associated with any one given target from a particular species where bacterial homologues are not available.

2. Solubilisation of membrane proteins

Membrane proteins by their nature are comprised of hydrophobic regions surrounded by lipid and hydrophilic regions in contact with the aqueous environment. Purification and subsequent crystallisation of any protein requires that protein to be in a soluble state. The solubilisation of membrane proteins involves the removal of the protein from the lipid environment whilst maintaining structural and functional integrity. For this to be achieved, the lipid molecules surrounding the protein must be exchanged for another factor which is also capable of protecting the hydrophobic regions from the surrounding solvent. The substance which is used to remove the membrane proteins from the lipid bilayer is detergent. There are many different detergents which are used to solubilise membrane proteins, but they are all used in the same way. Determining the detergent which is best suited for solubilisation of the target protein can be a complicated and tedious affair, mainly due to the intricate and complex relationship between the protein and the lipid bilayer.

2.1. DETERGENTS

Detergents have the same dual properties as lipid molecules and membrane proteins in that they have both hydrophobic tails and hydrophilic head groups. Detergents vary in both nature of the hydrophilic head group (sugar-based, phospholipid-like) and the length and composition of the hydrophobic alkyl tail (see Figure 3). At low concentrations the detergents exist as monomers in solution, however, at a specific concentration called the critical micelle concentration (CMC) the detergent molecules form micelles due to the hydrophobic effect (Figure 1). The CMC is inversely related to the length of the alkyl chain thus the longer the alkyl chain the smaller the value. However, the precise value is different for each different detergent and must be experimentally determined.

At the CMC, a detergent can effectively disrupt the interactions between a membrane protein and the lipid bilayer. In other words, this is the concentration at which a detergent breaks the membrane into detergent or lipid and detergent or protein micelles, and hence effectively solubilises a membrane protein. The detergent not only disrupts the interactions between the protein

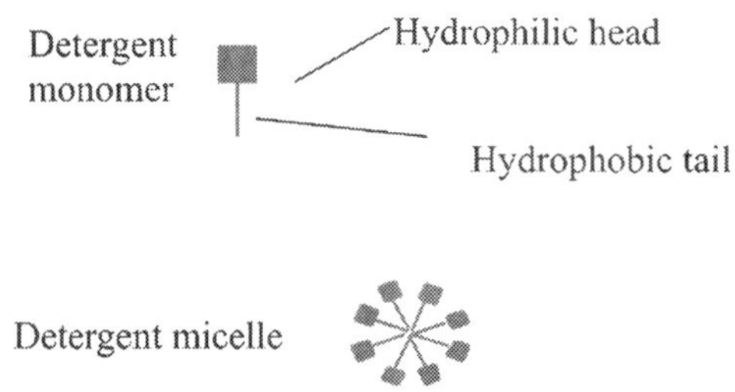

Figure 1. At low concentrations, detergent molecules exist as monomers. At a higher concentration, the critical micelle concentration (CMC) specific to each detergent, the detergent molecules spontaneously form micelles. In this state, the detergent can effectively solubilise lipid membranes.

and the lipid, but also effectively replaces the lipid because of its dual hydrophobic and hydrophilic nature (Figure 2). The hydrophobic tails of the detergent molecules associate with the hydrophobic portions of the protein, leaving the hydrophilic head groups in contact with the solvent.

The CMC is an important feature of detergents since this value can have major implications on how the solubilised protein is processed after purification (see detergent removal or exchange section). It is also important to keep in mind that the CMC varies with differing salt concentration and temperature, two important variables for many functional assays and crystallisation trials. The choice of detergent for a given membrane protein is often a case of trial and error; however, there are some general guidelines. Generally, membrane proteins are much happier and thus more stable in detergents with long alkyl chains such as dodecyl-B-D maltoside (DDM), a C12 detergent (Figure 3B), since these mimic the membrane better than those with shorter chains, e.g., octylglucoside (OG), a C8 detergent (Figure 3A). It is generally best to perform the initial solubilisation procedure using a detergent which will maintain the protein as stable as

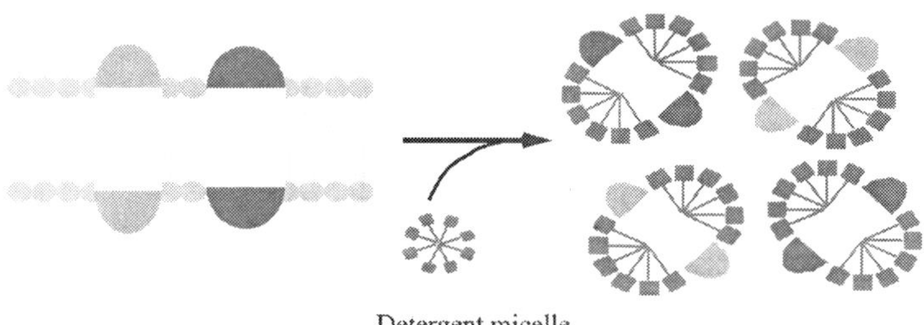

Figure 2. Addition of detergent at or above the critical micelle concentration (CMC) effectively disrupts interactions between the protein and lipid molecules in the membrane. The detergent replaces the lipid to generate protein–detergent micelles.

Figure 3. Structures of C8 and C12 detergents.

possible. For crystallisation as we shall see it is often necessary to exchange detergent.

The solubilisation procedure itself involves a simply incubating membrane or cell preparation with a solution of detergent above the CMC. The solution is usually incubated at 30 min at 4°C. A centrifugation step at 100,000 g for 30 min separates out the soluble from the insoluble material. The soluble protein fraction is then subjected to further purification procedures.

3. Purification

The procedures involved in the purification of membrane proteins are more or less the same as those used for soluble proteins. The major difference is that detergent must be present throughout the process in order to maintain the protein in a soluble state. In terms of the ultimate aim of structure determination there are two general ways in which it is possible to obtain protein sample suitable for crystallisation trials (see below for more details). Many of the available membrane protein structures are of targets which are expressed in large quantities naturally. In each case, it was necessary to develop individual multistep purification strategies which exploited the individual characteristics of the protein (e.g., formate dehydrogenase [7] and photosystem II [8]).

Due to their hydrophobic nature membrane proteins tend to form stronger interactions with the chromatographic media. This can lead to significant losses of protein. This is not a problem when large quantities of protein are readily available, however, can be devastating when the target protein is produced in small amounts. Limiting the number of purification steps can reduce these losses. This makes affinity purification, the so-called one-step purification method, an attractive alternative to traditional multistep purification methods. As previously mentioned, the most common affinity purification method employs the His tag and exploits the interaction between the histidine side chains and Ni^{2+}. Although in practice, purification of the target protein to homogeneity often requires a further step such as the use of an anion exchange column, the overall number of steps is usually reduced. The affinity chromatography methods have the added advantage of a standard protocol which although not suitable for all proteins at least provides a starting point for developing a strategy. There are now several examples of His-tagged membrane proteins which have yielded high-resolution structures, most notably lactose permease [9] and MsbA from *E. coli* [10]. These and other studies have also shown that although in some labs it is standard to remove His tags prior to crystallisation set-up, this does not always seem to be necessary.

4. Sample preparation

The eluate from the final step of the purification process needs to be concentrated and washed. Generally, it is best to exchange the protein into the simplest possible sample buffer (e.g., 0.1 M Tris pH 7.5 + detergent at the CMC). Once the protein has been effectively exchanged into the final buffer and is at a reasonable starting concentration for crystallisation trials it is often sensible to perform a number of quality control checks. The first and most important analysis is to run a small sample on an SDS-PAGE gel in order to check the purity of the final protein. We would aim for 85–95% purity of the sample; however, one can attempt crystallisation trials at much lower purities and obtain well-diffracting crystals [7]. One further useful step is to simply store a small amount of the pure protein at 4C and analyse aliquots on an SDS-PAGE gel over the course of 1 week or so. This will give an indication of the stability of the protein. Often what begins as a single band on the first day is observable as several bands after a number of days, indicating breakdown of the protein over time. It is possible to alter the buffer conditions to limit this breakdown.

It is also useful to check the aggregation status of the sample. As with soluble proteins we aim to obtain a fully monodispersed sample. The aggregation status of soluble proteins is usually assessed using dynamic light scattering (DLS), however due to the presence of detergent other methods have to be used with membrane proteins. One method uses single-particle electron microscopy This simply involves mounting a small amount of the purified soluble protein onto a carbon-coated grid, staining the protein, and examining it under an electron microscope. What we hope to see is a single population of the same-sized particles. However, what is often observed is a mixture of different-sized particles corresponding to the protein in various states of aggregation (Figure 4). It is possible to remove these large protein aggregates by high-speed centrifugation (10 k–150 k $\times$ g) for about 30 min.

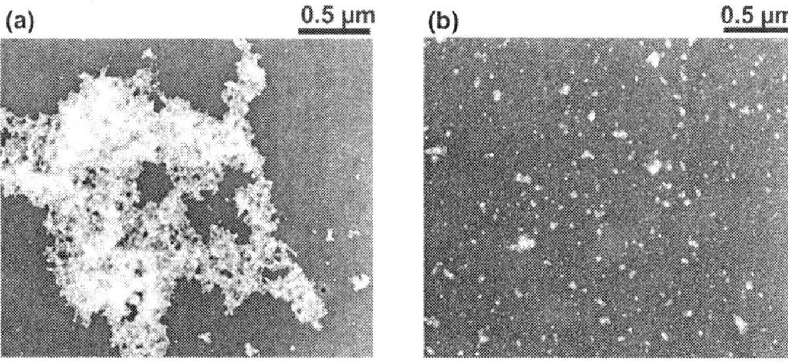

Figure 4. Electron micrographs of succinate–quinone oxidoreductase (SQR) protein sample before (a) and after (b) ultracentrifugation.

The heavier aggregates are removed in the form of a small pellet leaving a sample of protein which contains largely monomerically dispersed protein. Centrifugation time and speed are important variables which need to be optimised for each different protein and sometimes even for each different samples of a given protein. Centrifugation at too high speed or for too long may lead to loss of much of the protein in question, whereas centrifugation at too low speed or for too short a time may not effectively remove the protein aggregates. It is possible to further analyse the protein sample by electron microscopy after centrifugation in order to assess the status (Figure 4). An elegant way in which this technique was used is described in Horsefield et al. [11]. In this case, it had been possible to obtain crystals of the target protein succinate–quinone oxidoreductase (SQR) for a number of years. However, these were invariably small and diffracted poorly. Removal of the aggregates facilitated the growth of well-ordered crystals and led to the first reports of a high-resolution structure.

It is also possible to check the aggregation status of a sample using size-exclusion chromatography. Other analyses can also be performed on the protein prior to crystallisation set-up such as circular dichroism to check for correct folding and, if available, functional assay.

5. Crystallisation

Once a pure, monodispersed sample has been obtained it is possible to proceed with crystallisation trials. There are three different types of membrane protein crystal. The most common is known as the type II three-dimensional (3D) crystal represented in Figure 5. The growth of these crystals is similar

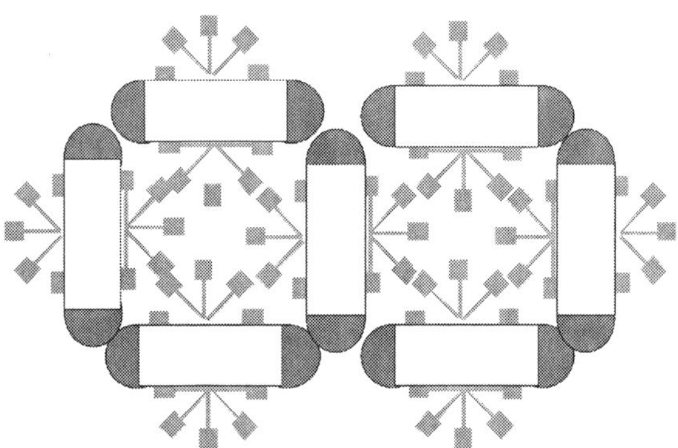

Figure 5. Type II 3D crystal. Detergent micelles cover the hydrophobic parts of the protein (white) leaving the hydrophilic regions (grey) exposed to the surrounding solvent.

to that of soluble crystals with the same techniques and precipitants used, however, the presence of the detergent has two impeding influences on crystal formation. As mentioned earlier, detergent micelles associate with the hydrophobic regions of the membrane protein leaving the hydrophilic regions exposed to the surrounding solvent. This limits the formation of the protein–protein contacts essential for crystal formation to the hydrophilic regions. In addition there is a finite space between the molecules which form the crystal lattice. In order for the crystal lattice to optimally form the detergent molecules must fit precisely within the gap between the molecules. These two factors mean that it is difficult to both obtain membrane protein crystals and even if obtained they are often small and diffract poorly.

One of the most critical factors therefore in the process of membrane protein crystallisation is screening of detergent. As mentioned earlier, membrane proteins tend to be more stable in detergents with longer alkyl chains. However, whilst these detergents are often excellent for initial solubilisation and maintaining the protein in a soluble state during purification, they can be unsuitable for crystallisation. Since they have larger micelles and cover more of the protein, they limit the formation of protein–protein contacts. It is therefore critical to change the detergent to expose more of the protein molecules. This can be performed in a number of ways. One detergent can be completely exchanged for another either through a molecular weight cut-off filter or more efficiently by detergent exchange on a chromatography column. Rather than complete exchange, it is also possible to crystallise in mixed micelles, where a mixture of two or more detergents of differing alkyl chain lengths combine to provide a stable environment whilst exposing larger regions of the protein molecules. This type of approach was first pioneered by Hartmut Michel for the crystallisation of the reaction centre of *Pseudomonas viridis*, the first ever membrane protein structure solved. However in this case, rather than using a mixture of detergents Michel's group used a small amphiphile, heptanetriol, a molecule with the same properties as a detergent, as an additive to the crystallisation drop [12]. The same amphiphile was also used for the successful crystallisation of bovine rhodopsin [13].

6. Antibody fragments

The selective use of detergents can alter the exposed region of the protein of interest. An alternative approach aims to increase the hydrophilic domain of a membrane protein by the addition of a specific antibody fragment. The first use of this approach for a membrane protein was also pioneered by Hartmut Michel's lab and resulted in the structure of cytochrome c oxidase from *Paracoccus denitrificans* [14]. In this case, it proved impossible to obtain crystals of the protein in dodecylmaltoside, the only detergent which maintained it in a stable state. The addition of the antibody fragment, an Fv fragment, extended the hydrophilic domain and allowed the formation of crystals [15]. All the protein–protein contacts within the crystal lattice are mediated by the antibody fragment.

The production of antibody fragments as protein-specific epitopes for structural studies is seldom routine for labs due to the long lead time and relative expense. In summary, the process involves immunising mice against the protein of interest. Antibody-producing B cells from the spleens of the immunised mice are fused with immortalised myeloma cells to produce hybridoma cells. Since each individual B cell from the spleen produces an individual antibody in this way it is possible to develop cell lines which produce a single, monoclonal antibody, essential for structural studies. The antibodies by their nature are secreted from the hybridoma cell into the surrounding tissue culture medium and it is this media which must be harvested and tested for the presence of protein-specific antibodies. It should be mentioned that the process involved in generating these hybridoma cell lines from immunisation to final screening can take between a few months and more than 1 year.

Whole antibodies are not good for cocrystallisation as they are bivalent, large, and have high internal flexibility due to their hinge regions. The antibody produced by the hybridoma cell can be used in two ways to facilitate the crystallisation and/or increase resolution of a membrane protein structure. The first method involves the generation of Fab fragments. Papain digestion is used to cleave the Fab fragment from the Fc (fragment crystallisable) fragment (Figure 6). The two can then be separated by ion exchange chromatography. The advantage of this method is that it allows more rapid generation of the antibody fragment than by attempting expression of an Fv fragment. However, it is more expensive to produce large quantities of antibody from a mammalian cell-based expression system. Furthermore, the digestion can lead to heterogeneous products which may be difficult to separate and characterise. This method has been successfully applied to the crystallisation of a K + channel [16]. In this case, the addition of the Fab fragment improved resolution of the protein from 3.2 to 2 Å.

Fv fragments are a much more popular choice as structural epitopes with many more examples of successful application to the crystallisation of membrane proteins in the literature. The Fv fragment is a globular and monovalent domain of roughly 25 kDa that can be recombinantly produced in *E. coli*. The starting point is again the hybridoma cell line. The hybridoma cell line contains the DNA coding for the antibody and it is this DNA which is required in order to make an expression construct for the production of the Fv fragments (Figure 6). The Fv fragment can then be produced to high

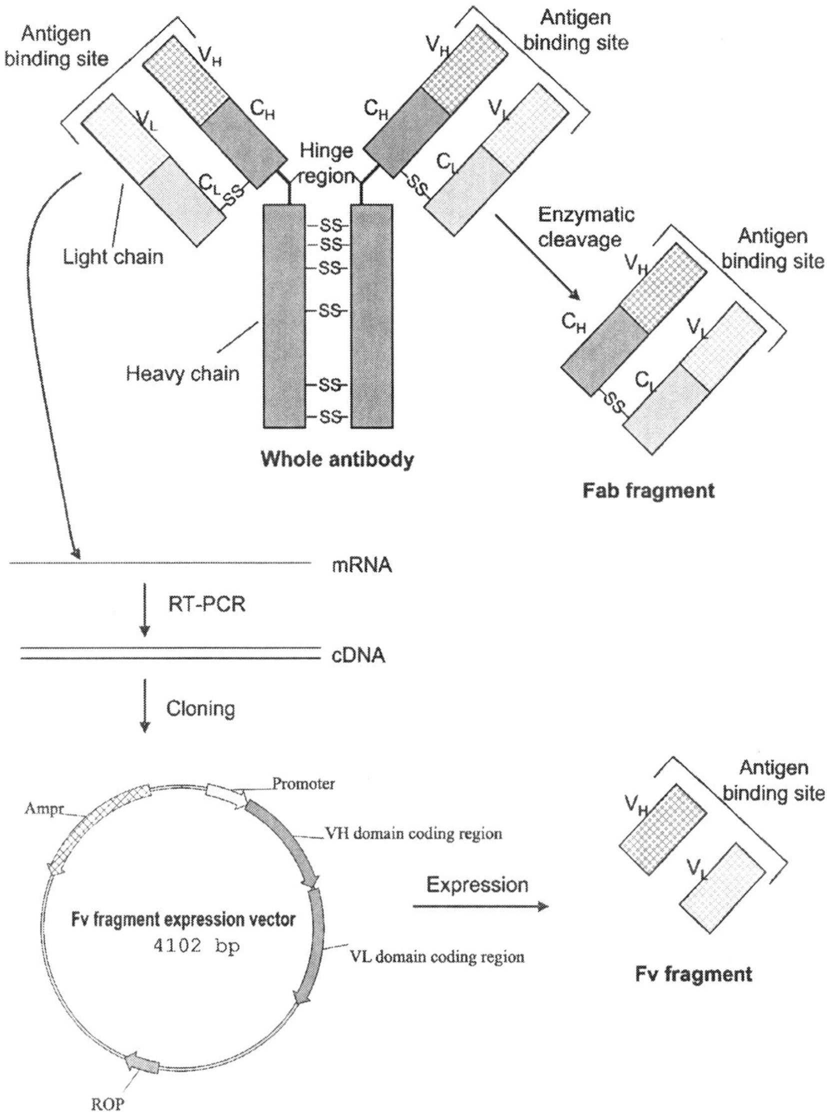

Figure 6. Summary of the production of both Fab and Fv fragments for structural studies.

levels, purified and mixed with the protein of interest prior to crystallisation trials. Along with cytochrome *c* oxidase this approach has been successfully applied to the crystallisation of cytochrome bc_1 complex from yeast [17].

Clearly, antibody fragments have a role in the structure determination of membrane proteins. However, it seems that they are best used as an optimisation tool when crystals of a target protein have been obtained. For most labs the investment in terms of time and money is simply too great to risk antibody production for a protein which has yet to yield diffracting crystals.

7.   Summary

Membrane protein structure determination has come a long way in the last 5 years or so with the number of high-resolution structures rising almost exponentially. This increase in knowledge in the areas of expression, purification, and crystallisation has allowed the development of more rational approaches and made this area more accessible to new researchers. It is anticipated that the next 10 years will see major breakthroughs, particularly in the area of medically relevant eukaryotic membrane proteins.

References

1. Kunji, E.R.S., Slotboom, D.-J., and Poolman, B. (2003) *Lactococcus lactis* as host for overproduction of functional membrane proteins. *Biochimica et Biophysica Acta*, **1610**: 97–108.
2. Long, B.S., Campbell, E.B., and MacKinnon, R. (2005) Crystal structure of a mammalian voltage-dependent Shaker family K + channel. *Science*, **306**: 809–902.
3. Bannwarth, M. and Schulz, G.E. (2003) The expression of outer membrane proteins for crystallization. *Biochimica et Biophysica Acta*, **1610**: 37–45.
4. Buchanan, S.K. (1999) B-barrel proteins from bacterial outer membranes: structure, function and refolding. *Current Opinion in Structural Biology*, **9**: 455–461.
5. Keifer, H. (2003) In vitro folding of alpha-helical membrane proteins. *Biochimica et Biophysica Acta*, **1610**:57–62.
6. Van den Berg, B., Clemons, W.M. Jr., Collinson, I., Modis, Y., Hartmann, E., Harrison, S.C., and Rapaport, T.A. (2004) X-ray structure of a protein conducting channel. *Nature*, **427**: 36–44.
7. Jormakka, M., Tornroth, S., Abramson, J., Byrne, B., and Iwata, S. (2002) Purification and crystallization of the respiratory complex formate dehydrogenase-N from *Escherichia coli*. *Acta Crystallographica*, **D58**: 160–162.
8. Ferreira, K.N., Iverson, T.M., Maghlaoui, K., Barber, J., and Iwata, S. (2004) Architecture of the photosynthetic oxygen-evolving center. *Science*, **303**: 1831–1838.
9. Abramson, J., Smirnova, I., Kasho, V., Verner, G., Kaback, H.R., and Iwata, S. (2003) Structure and mechanism of the lactose permease of *Escherichia coli*. *Science*, **301**: 610–615.
10. Chang, G. and Roth, C.B. (2001) Structure of MsbA from *E. coli*: a homolog of the multidrug resistance ATP binding cassette (ABC) transporters. *Science*, **293**: 1793–1800.
11. Horsefield, R., Yankovskaya, V., Tornroth, S., Luna-Chavez, C., Stambouli, E., Barber, J., Byrne, B., Cecchini, G., and Iwata, S. (2003) Using rational screening and electron microscopy to optimize the crystallization of succinate:ubiquinone oxidoreductase from *Escherichia coli*. *Acta Crystallographica Section D: Biological Crystallography*, **59**: 600–602.
12. Deisenhofer, J., Epp, O., Miki, K., Huber, R., and Michel, H. (1984) X-ray structure analysis of a membrane protein complex. Electron density map at 3 A resolution and a model of the chromophores of the photosynthetic reaction center from *Rhodopseudomonas viridis*. *Journal of Molecular Biology*, **180**: 385–398.

13. Okada, T., Le Trong, I., Fox, B.A., Behnke, C.A., Stenkamp, R.E., and Palczewski, K. (2000) X-ray diffraction analysis of three-dimensional crystals of bovine rhodopsin obtained from mixed micelles. *Journal of Structural Biology*, **130**: 73–80.
14. Ostermeier, C., Harrenga, A., Ermler, U., and Michel, H. (1997) Structure at 2.7 Å resolution of the *Paracoccus denitrificans* two-subunit cytochrome c oxidase complexed with an antibody FV fragment. *Proceedings of the National Academy of Sciences of the USA*, **94**: 10547– 10553.
15. Ostermeier, C., Iwata, S., Ludwig, B., and Michel, H. (1995): Fv fragment-mediated crystallization of the membrane protein bacterial cytochrome *c* oxidase. *Nature Structural Biology*, **2**: 842–846.
16. Zhou, Y., Morais-Cabral, J.-H., Kaufman, A., and MacKinnon, R. (2001) Chemistry of ion coordination and hydration revealed by a K + channel–Fab complex at 2.0 Å resolution. *Nature*, **414**: 43–48.
17. Hunte, C., Koepke, J., Lange, C., Roßmanith, T., and Michel H. (2000) Structure at 2.3 Å resolution of the cytochrome bc_1 complex from the yeast *Saccharomyces cervisiae* co-crystallized with an antibody Fv fragment. *Structure*, **8**: 669–684.

MACROMOLECULAR CRYO-CRYSTALLOGRAPHY

ELSPETH GARMAN

*Laboratory of Molecular Biophysics, Department of
Biochemistry, University of Oxford, OX1 3QU, UK*

Abstract: During room temperature X-ray data collection, macromolecular
crystals commonly suffer severe radiation damage. Thus, diffraction data are
now routinely collected with the sample held at around 100 K, significantly
reducing the secondary radiation damage, and usually resulting in higher
resolution and better quality data. However, at synchrotron X-ray sources,
even at cryo-temperatures there has now been frequent observation of both
degradation of data quality as the experiment proceeds and specific structural damage to particular amino acids due to radiation damage. Present
research into cryo-techniques seeks to understand the basic physical and
chemical processes involved in flash-cooling and radiation damage, to allow
rational optimisation of cryo-protocols and minimisation of the deleterious
effects of X-ray irradiation.

Keywords: cryo-crystallography; cryoprotection; radiation damage; mosaicity;
secondary radicals; ice.

1. Introduction

Many proteins and nearly all virus crystals are extremely susceptible to radiation
damage and only diffract for a few minutes in the X-ray beams now provided by
second and third generation synchrotrons. In fact, by the early to mid-1990s,
radiation damage was becoming a limiting problem in the full utilisation of
newly available intense synchrotron sources. Radiation damage is caused initially
by "primary" interactions between the atoms in the crystal and the beam, during which the X-rays lose energy by the photoelectric effect and by inelastic
(Compton) scattering. The extent of this primary damage is *dose* dependent.

 This energy is dissipated in several ways: it produces heat (thermal vibration
of the molecules) and it provides the energy to the photoelectrons both to
ionise other atoms and to result in the breaking of bonds between the atoms
in the molecules. It can produce reactive radicals by two predominant

25

R.J. Read and J.L. Sussman (eds.): Evolving Methods for Macromolecular Crystallography, 25–40
© 2007. *Springer*

mechanisms: (1) direct (e.g., damage to polypeptide) and (2) indirect (reactive H* or OH* produced by destruction of a water molecule). Any thermal energy will allow these reactive products to diffuse through the crystal causing further destruction ("secondary" damage). This component of the radiation damage is *time* and *temperature* dependent.

At room temperature, the reactive products diffuse and spread through the crystal causing further damage as they go, and this time-dependent secondary damage, in addition to the primary dose-dependent part, often destroys the crystal. At cryo-temperatures (around 100 K) most of the reactive products are immobilised in the crystal and do not cause extensive secondary damage in areas of the crystal that are not exposed to the beam [1–6]. Thus, cooling the samples for data collection significantly reduces, but *does not eliminate* radiation damage, since the primary damage will always occur and some secondary products, particularly electrons, are known to be still mobile at cryo-temperatures [7].

In fact, three systematic studies published in 2000 [8–10] on the effects of radiation damage on molecular structure showed that structurally specific effects are evident in the electron density maps calculated from sequentially collected cryo-data sets. The most notable of these is the breaking of disulphide bridges. A range of further experiments are now underway to understand the damage processes and design data collection strategies to minimise them [11, 12].

Because protein crystals usually have a high water content, cooling them to around 100 K without compromising their diffraction quality requires particular techniques. Work by Hope [13] in extending small molecule cryo-techniques for use with protein crystals, prompted a number of researchers to undertake serious experimentation. Currently, the simplest and most generally used technique is the loop mounting method of Teng [14], which has been developed further since it was originally reported in 1990.

2. Historical notes

- 1958. Ribonuclease II cooled in methyl-2,4-pentanediol (MPD) to −27°C. Resolution improved from 1.69 to 1.41 Å [15].

- 1966. Orthorhombic insulin + heavy atom cation at −150°C [16]. Mosaic spread increased to a range of 0.5–1.0°C.

- 1968. Orthorhombic lysozyme at −100°C [17]. Cross-linked with gluteraldehyde and soaked in 50% glycerol solution. Crystals were mounted in glass capillaries and cooled by boiling liquid nitrogen through a glass coaxial with the capillary. Worked well.

- 1970. Dogfish lactate dehydrogenase at –50°C [18]. Method of (c) destroyed crystals. Soaked in 3 M sucrose-ammonium sulphate solution prior to dipping in liquid nitrogen. Made a "glass", that did not expand so did not crack crystal. Crystal was mounted on a GLASS FIBRE. Problem was that unit cell changed with sucrose concentration, BUT 10× less radiation damage than at room temperature.

- 1973. Sperm whale myoglobin at 77 K with a hydrostatic pressure of 2,500 atm to form ice phase III phase [19]. Same diffraction observed as at room temperature. This idea has been revived very recently and shows some promise [20].

- 1975. General method proposed by Petsko [21]: replace normal mother liquor with salt-free aqueous organic solvent of low freezing point, e.g., MPD, isopropanol, ethylene glycol. Have to find right one for your protein.

- 1988. Method of Hope [13], which had already been used for small molecules for 10 years, was first applied to a protein (crambin [22]). Crystals were covered in oil, and then cryo-cooled on a very thin glass spatula. Variations on this theme followed, and some protein data were collected.

- 1990. Major breakthrough with method of Teng [14], where crystal is mounted in a small loop and supported by the surface tension of a cryoprotected solution. Since then the technique has flowered, and its use is now routine and extremely widespread.

- 1990–present. Loop mounting macromolecular cryo-techniques for flash-cryo-cooling, storage, and retrieval of crystals developed and become the accepted method of data collection. The techniques are reviewed in detail in [1–4, 6, 23] to which the reader is referred for much fuller descriptions.

3. Crystal treatment

The crystal is first soaked for anything between 1 s and a few days (but usually around 3 min) in a "cryosolution". The crystal is then held by the surface tension of the cryo-solution across a loop made of thin fibre and is immediately plunged into a cryogen such as gaseous nitrogen at around 100 K, or liquid nitrogen, propane or ethane. The film of cryo-solution becomes solid and holds the crystal rigidly in the loop. The antifreeze in the cryo-solution and the rapid rate of freezing enable the crystal to be cooled to cryo-temperatures with no ordered ice formation; instead, a vitreous glass is formed, which does not disrupt the crystal order or interfere with the diffraction.

The first step in preparing to flash-cool, a protein crystal is to find a suitable cryo-solution. This is usually the mother liquor with a cryoprotectant agent ("antifreeze") added (e.g., glycerol, ethylene glycol, MPD, light polyethylene glycols [PEGs], various sugars) in high enough concentration to prevent the formation of ordered ice and to promote vitrification of all the liquid in the sample. The mother liquor should not be diluted by the cryo-agent: the cryo-solution should be made up so that water in the mother liquor is replaced by cryo-agent. A test with the putative cryo-solution flash-cooled in the loop without the crystal will give the correct concentration needed. A drop of the chosen cryoprotected mother liquor alone, placed in the loop with a low volume pipette, should first be flash-cooled to check that it forms a transparent glass. A drop, rather than a film, should be used, as this better simulates the presence of a crystal. A clear drop in the loop is a necessary but not sufficient criterion: a diffraction image should always be taken to check for diffuse rings from ice crystallites. Figure 1 shows the diffraction images obtained from a mixture of water and glycerol cooled in a mohair loop (water/glycerol v/v): 100%/0%, 95%/5%, 90%/10%, 80%/20%, 70%/30%, 65%/35%, and 60%/40%, and the resolutions at which ice rings appear.

Satisfactory cryoprotection is afforded by addition of 40% glycerol, where the diffuse scattering ring has similar slope on the high and low resolution side [6]. The temptation is to stop when an image like 65%/35% is obtained, when in fact an image like the 40%/60% one will usually give better results (lower mosaic spread and better diffraction). If transferring a crystal and mother liquor to the cryo-solution, the cryo-solution will be diluted, and thus it is wise to increase the cryo-agent concentration by 2–5% as a safety margin, to take this dilution into account.

It is also worth testing the diffraction of a crystal in cryoprotectant at room temperature, since it is important to determine if the addition of cryo-protectant or the flash-freezing itself is the cause of a badly diffracting or a

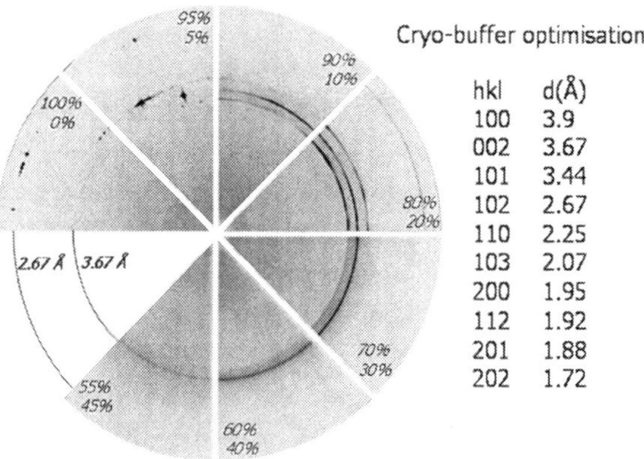

Figure 1. X-ray diffraction patterns of water/glycerol (v/v) mixes, and the resolutions at which ice rings appear [6]. (Reproduced from [6]. With permission of the IUCr.)

mosaic crystal. A very convenient room temperature loop mounting method can be found in reference [24] which is easier to execute successfully than the traditional glass capillary technique.

Next, the crystal is introduced into the cryo-solution broadly in three different ways. For most crystals, the easiest option, which is usually satisfactory, is to transfer the sample from its growing drop straight into the final concentration of cryo-solution and leave it for 30 s–4 min. It must be emphasised that this soak time is rather empirical; a longer soak allows more time for equilibration but prolongs the time for which the cryo-solution can degrade the crystal, whereas a short soak maximises the osmotic shock to the crystal but minimises the degradation time. Times from 0.5 s (crystal dragged through cryo-solution) to several days have been used successfully, and there is a huge variation in soaking times used in different laboratories practising cryo-crystallography. Sequential soaks in increasing concentrations can also be used, and these lessen the osmotic shock to the crystal (see Sections 7 and 8.4).

Cryoprotectant agents can also be dialysed into the crystal, and this method often succeeds where soaking has failed, see for example [25, 26]. However, the ideal case, which minimises the crystal handling, is where the crystals are grown in a mother liquor which is already adequately cryoprotected. In fact, a number of crystal screens are now commercially available. An account of this approach is given in [27].

Nearly, all MX laboratories are now equipped to allow rapid mounting of the crystal in the loop from the cryo-solution onto the goniostat (see Figure 2). A modified goniometer head with a small disc magnet with a bevelled locating pin attached in the centre can be used. Some sort of arrangement like this is essential for the speedy transfer of crystals. The locating pin fits into a hole in the bottom of a metal button ("top-hat") into which the hollow pin and loop are clamped.

Usually, the loops should be matched to the size of the crystal, as then the crystal is easier to locate when cooled. They can be made of any thin (20–50 μm thick) fibre, but can be purchased as pre-made in rayon and thin nylon. Recent developments have provided experimenters with two more robust alternatives to the commercially available rayon loops. These are the "litho-loops" which are etched from mylar film and come in a range of circular and elliptical sizes, and the "micromounts" [28] which are fabricated from patterned and shaped thin polyimide film. Both of these represent significant improvements, since they do not unwind with time, they are precisely the size they claim to be and they are more rigid than fibre loops.

The volume of cryo-solvent around the crystal can be minimised by lifting the loop out of the cryo-solvent so that its plane is perpendicular to the surface of the drop (see Figure 3a).

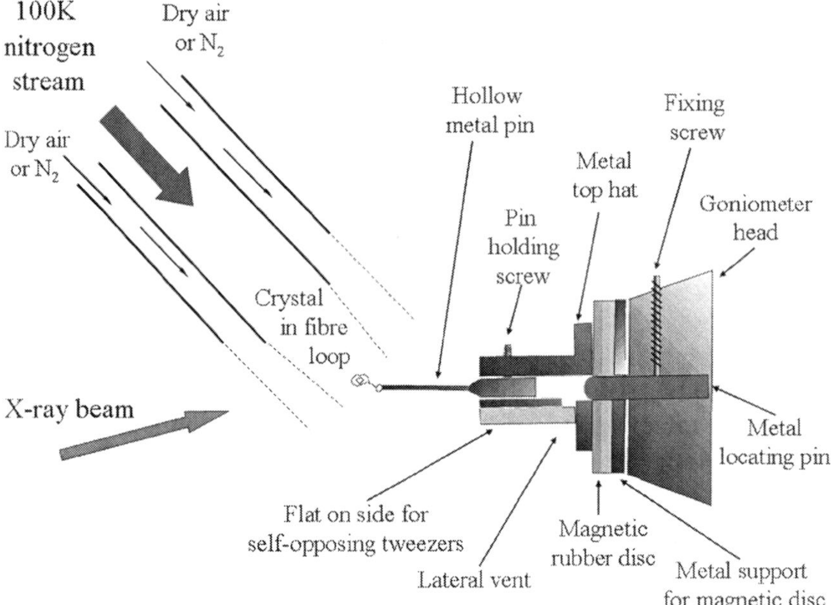

Figure 2. A typical experimental arrangement for a cryo-crystallographic data collection. (Reproduced from [1]. With permission of the IUCr.)

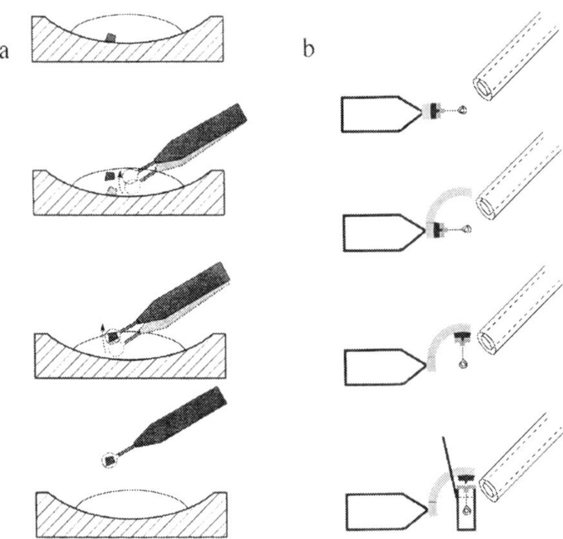

Figure 3. (a) Method of fishing a crystal to minimise thickness of cryo-solution thin film [27]. (b) Use of removable arc to store a cryo-cooled crystal for later use [1]. (Reproduced from [1]. With permission of the IUCr.)

4. Storage of cryo-cooled crystals

Crystals can be cooled in either a gaseous nitrogen stream or straight into liquid cryogen. There is some debate as to which method achieves the fastest cooling (see Section 8.2). Measurements on cooling rates are notoriously difficult to make accurately, since the device used to monitor the temperature actually takes heat into the system.

There are several advantages of being able to store cooled crystals: (1) crystals can be screened for quality on a home source prior to synchrotron data collection; (2) crystals can be cooled when in peak condition for later use (old crystals often do not diffract as well as fresh ones); and (3) extra data can be collected at a later date if the data collection is interrupted for any reason. A cooled crystal can be stored for future use in a cryo-vial of nitrogen.

The techniques for storing crystals [1, 2, 23] are still evolving and it is hoped that the push towards automation at a number of sites will result in better and more reliable methods becoming routine in the near future.

With careful design of the geometry of the experiment, and by using a removable arc goniometer or a simpler "flipper" the crystal can be moved so that the loop points downwards (see Figure 3b). (*Note*: the crystal must remain in the nitrogen stream at *all* times while it is rotated.) Once in this position, the vial of liquid nitrogen or propane can be brought underneath it. Alternatively, pre-cooled especially designed "Hope tongs" can be used to transfer the crystal between liquid nitrogen and the nitrogen gas stream. However, when using the tongs, great care must be exercised not to block the cryostat stream with them when approaching the crystal. As regards cryogen, the use of propane rather than nitrogen has now almost died out due to safety concerns regarding its transport on commercial aircraft.

5. Heavy atom and substrate soaks

When performing heavy atom, substrate, and inhibitor soaks, the heavy atom should be added to the cryo-solution, even if the cryo-soak is short. Occasionally competitive binding will occur between the cryoprotecting agent and the desired substrate (e.g., phosphorylase $\underline{b}$ with glycerol [29], and a new, non-competitive cryoprotecting agent will have to be found (e.g., for phosphorylase, MPD was used successfully). Bound cryo-agent molecules are also commonly observed in electron density calculated from cryo-data sets.

When undertaking substrate binding or inhibitor studies using data from cooled crystals where the native model is from room temperature data, it is *essential* to refine the data to the native model (excluding substrate) *before* calculating a difference map. This is because cell shrinkage on freezing can be significant, and noise from this can obliterate the signal from a bound substrate. Note that a 0.5% change in all the cell dimensions of a 100 Å unit cell will result in an average change of approximately 15% in reflection intensities within a 3 Å sphere [30].

6. Ice

In spite of efforts to avoid ice formation (see [1] for detailed discussion), some ice may gradually accumulate on the sample during the experiment. This will be evident on the diffraction pattern, either as diffuse rings, sharp rings or individual reflections (see Figure 1). Ice can be carefully removed with a small artist's brush or with an acupuncture needle attached to a non-heat-conducting rod. Alternatively, a little liquid nitrogen can be poured onto the loop, taking care not to damage any surrounding equipment (e.g., camera lens beneath).

However, if there are ice features on the images, their effect on the data quality can be minimised during data processing. If the software allows, the ice diffraction is best masked out before integration rather than at the scaling stage. The same software commands can be used to mask out reflections from salt crystals which appear at lower resolution than those from ice (see Figure 1). Salt crystals sometimes form in high salt cryo-solutions during the transfer of the loop to the goniostat.

Monitoring the images as they are collected and removing ice if it forms is the best way to minimise the problem.

7. Mosaicity

A common observation is that the mosaic spread of cryo-cooled crystals tends to be higher than that of the same crystal at room temperature. Increased mosaicity adversely affects the data quality since it increases the number of overlaps, it decreases the signal-to-noise ratio of a reflection as the reflection is spread over a larger volume of reciprocal space, and for two-dimensional (2D) data collection, it decreases the ratio of fully to partially recorded reflections. The decrease in signal-to-noise ratio with increasing mosaicity usually means that if the mosaicity is minimised, the diffraction limit is maximised (see [31] where this was shown for crystals of phosphorylase b).

Handling of the crystals can cause increased mosaic spread; handling should thus be minimised. By investing some time in exploring different cryo-conditions, cryo-solutions, and soaking methods, it should be possible to reproduce room temperature mosaicity in most cases. Rocking curves can most easily be collected and investigated on an electronic area detector (a charge-coupled device [CCD]-based detector) where three-dimensional (3D) data collection is time efficient.

Sequential soaking in increasing concentrations of cryo-solution, rather than putting the crystal straight into the final concentration, can reduce the mosaicity. During a sequential soak, both the handling of the crystal and the severity of the osmotic shock to it of the 5% or 10% increasing concentrations of cryo-agent, can be lessened by the following simple measure. The crystal and mother liquor is left in the same microbridge or soaking well and a 10% cryo-solution is pipetted onto it, the drop is agitated with the pipette end without touching the crystal, and then some of the liquid (say 10 µl) is removed. Ten microlitres of the 10% solution is now added again, mixed, 10 µl taken off, 20% cryo-solution added, mixed, and so on until the desired concentration of cryo-solution is reached [3].

For example, for crystals of the 42 kD neuraminidase from *Salmonella typhimurium*, time was spent experimenting with cryo-protocols to optimise diffraction. The above method resulted in routinely obtaining resolutions better than 1 Å at the synchrotron, whereas moving the crystal between microbridges during the sequential soak gave resolutions worse than 1 Å, and non-sequential soaking gave around 1.3 Å.

Several factors can affect the mosaic spread during the actual cooling procedure. Swift transfer from the drop to the liquid cryogen or goniostat is desirable, since the crystal surface can dehydrate while travelling through the air. Placing the microscope as close as possible to the cryogen or goniostat, as well as practising the transfer on several "dry runs" beforehand, minimises the transfer time.

The rate of crystal cooling is another important factor: in general, the faster the cooling the lower the resulting mosaic spread. Since the rate of cooling is critically dependent on the crystal surface to volume (S/V) ratio, those with larger S/V (and thus larger surface area/thermal mass) often cool better than those with smaller S/V (cf. insects having high S/V which do not survive in the winter). Experience has shown that crystals with around 0.4 mm in their largest dimension and a surface to volume ratio of more than 12 mm^{-1} tend to cool better. However, a balance must be struck between the desirable increase in diffracting power of high volume crystals and the undesirable increase in mosaicity observed when flash-cooling them.

The choice of cryogen can also affect the rate of cooling and hence the mosaic spread (see Section 8.2). Crystal annealing can also reduce the mosaic spread (see Section 8.5).

8.   If nothing seems to work

Occasionally the experimenter is faced with a problem protein crystal, for which an appropriate cryo-solution cannot be found, or having found a seemingly benign cryo-solution, the diffraction is still not satisfactory. In these circumstances, there are various aspects of the cryo-cooling procedure that can be changed in a search for better results. However, it is always worth having more than one attempt at flash-cooling before giving up on a particular set of conditions, as small differences in procedure can often make the difference between success and failure.

Before embarking on any of the experiments suggested below, it is very useful to collect some data from a crystal at room temperature (see above) to give an idea of its intrinsic diffraction power and mosaicity. If there is no diffraction beyond 6 Å, or the crystals are highly mosaic, it is unlikely that cryo-cooling will improve them. Having established, for instance, that the crystals diffract to 4 Å on an in-house source at room temperature, it is worth the effort of finding appropriate cryo-conditions.

8.1. CRYO-SOLUTIONS

If the cryo-solutions have been made up by adding cryo-agents to the original mother liquor stock solution, it is worth making up a more concentrated stock solution of mother liquor and adding cryoprotecting agents to this in such proportions that the original mother liquor component concentrations remain the same, i.e., the cryoprotecting agent replaces water in the mother liquor rather than diluting it. For instance, if the mother liquor has concentration X, make up a 2X solution, and to make 100 µl of 5% cryo-solution, add 50 µl 2X, 5 µl glycerol, and 45 µl water.

As well as the more common cryoprotecting agents already mentioned, others such as erythritol, xylitol, inositol, raffinose, trehalose, 2,3,R,R-butanediol, propylene glycol, isopropanol (concentrations up to 70% required), DMSO (dimethyl sulphoxide), and other alcohols have been used successfully. This list is by no means exhaustive and mixtures of two different agents have also been found helpful, as well as treatment with a cryo-solution followed by immersion in oil such as Paratone N [32] or various silicone oils [33] prior to flash-cooling. Crystals have also been found to tolerate flash-freezing better after being gently cross-linked with gluteraldehyde [34].

If no benign cryo-solution can be found, another option is to exchange the mother liquor in the crystals for another solution in which the crystals are stable and which can be more easily cryoprotected. This exchange may have to be performed slowly and/or gradually. A convenient vapour diffusion method of finding an alternative solvent has been reported [35]. If the crystals do not react well to a sudden change of solvent, a flow cell can be used. For instance, in the study of phosphoglucomutase, 2 M ammonium sulphate and PEG 3350 was replaced over many hours in a flow cell by PEG 600 [36]. A decrease in mosaic spread and an improvement of the diffraction limit from 2.75 to 2.35 Å were observed.

8.2. CRYOGEN CHOICE

The most common cryogens in use are liquid nitrogen (melting point [MP]: 63 K, boiling point [BP]: 78 K), propane (MP: 86 K, BP: 231 K), and gaseous nitrogen. Generally, the simplest technique is to stream cool into gaseous nitrogen. Measurements of the cooling rates of various cryogens are notoriously difficult to make. For samples comparable with a typical protein crystal, there are two reported sets of measurements [37, 38]. Both concluded that gaseous nitrogen is the slowest in cooling between 300 and 100 K, but were at variance on whether liquid propane or liquid nitrogen was faster overall. The results depend critically on the size of the sample and the cooling regimes (nucleate or film boiling). If stream freezing has failed, both propane and liquid nitrogen are worth a try.

Note that there are safety implications of using propane in the laboratory, and now great problems may be experienced in shipping dry Dewars with crystals cooled in propane to synchrotrons. More experimenters are now also trying ethane as a cryogen, and freon 12 and methylcyclopentane have also been used.

A recent theoretical study modelling cooling rates for the flash-cooling of protein crystals in loops [39] concluded that the choice of cryogen was of relatively low importance to successful cryo-cooling. The crystal solvent content and solvent composition came top of the list, followed by the crystal size and shape (crystals with large surface to volume values cool faster and more uniformly than those with small S/V), amount of residual liquid around the crystal (which should be minimised), the cooling method (liquid or stream cooling), choice of gas or liquid cryogen, and lastly the relative speed between the cooling agent and the crystal. This theoretical study has given some rationale to procedures which have been empirically determined over the last 10 years.

8.3. TRANSFER, HANDLING, AND SOAKING PROCEDURES

Some protein crystals are very sensitive to any handling, and also the way the cryoprotectant agent is introduced can have a great impact on the observed diffraction pattern. As a general rule, handling should be minimised. It is worth thinking about the way the crystals are treated from the moment of opening up the tray in which they were grown to the point they are flash-cooled, and considering all the stages where degradation of crystal order might occur. For manipulation of crystals, acupuncture needles are extremely useful and are also cheap. Loops can be used to move crystals gently from the growing drop to the soaking well. For sequential soaks, crystals are better not moved between soaking wells (see Section 7).

Careful vapour pressure equilibration is very well worth trying. The cover slip with the hanging drop containing the crystals can be sealed over a volume of cryo-buffer in the bottom of a tray overnight, to allow equilibration of the vapour components. The crystal is then given a quick soak as before. This method has been found to be less invasive by some researchers, who always use it (Steve Gamblin, 2005, private communication).

Another variable which can be explored is the temperature of the cryo-solution soak, e.g., leaving the crystals in cryo-solutions overnight at 4°C prior to flash-cooling.

A critical step in the cryo-cooling procedure is the time taken to transfer the crystal from the soaking drop to the cryogen; this should be as swift as possible. If stream freezing straight onto the goniostat, it is strongly advisable to block the stream with a narrow piece of card until the crystal is safely positioned in the pre-centered place, and the transfer tweezers are well out of the way. The card is then quickly removed. This ensures the crystal is truly "flash-cooled", does not suffer dehydration in the dry room temperature air or nitrogen stream, and also prevents inadvertent knocking of the crystal out of the stream after it has been cooled.

8.4. OSMOLARITY MATCHING

When they are soaked in cryo-solutions, crystals often suffer from a large osmotic shock and are thus compressed, resulting in cracks, mosaic spread increase, and resolution degradation. One approach to overcome this is to match the osmolarity (Os/l) or osmality (Os/kg) of the mother liquor (or stabilising solution) and cryo-solution by modifying the concentration of the stabilising solution [3]. Osmalities are tabulated for most of the commonly used solutions in [40].

8.5. CRYSTAL ANNEALING

Some researchers have had success in extending the diffraction limits and decreasing the mosaicity of their crystals by the technique of crystal annealing, where the cooled sample is allowed to thaw to room temperature and is then flash-cooled again, sometimes being cycled in this way several times. Two methods of cycling have been reported. In one [41], the crystal was rapidly thawed and recooled in situ on the goniostat by blocking the gas stream for 1.5–2 s and waiting 6 s before repeating the process twice more. In the other [42, 43], the cooled crystals were removed from the goniostat and replaced into the cryo-solution for at least 3 min before being cooled again.

Annealing is sometimes spectacularly successful, but certainly not always. Thus, both methods of annealing outlined above are worth trying, unless there is only one crystal: in this scenario, some data should be collected first in case annealing makes the diffraction worse. It has been found that warming hen-egg white lysozyme (HEWL) crystals up to 230–250 K and then flash-cooling them again gave more reproducible results than did warming up to 293 K, and this may be a generally applicable strategy. From these experiments, understanding of the annealing process is now starting to emerge [44]. In addition, Juers and Mathews have shown that bulk solvent leaves the crystal during annealing if the cryoprotectant agent concentration is below the optimum (defined as giving the best diffraction properties), and that water enters the crystal during annealing if the initial cryoprotectant concentration is higher than the optimum. The experimenters concluded that during annealing, the cryoprotectant agent concentration adjusts itself, thereby changing its thermal properties so that the bulk solvent contraction when cooled more nearly matches the contraction of the crystal lattice. They further found that annealing is more likely to be successful if the initial concentration of cryoprotectant agent is above, rather than below the optimum [45]. These results imply that it is better to err on the side of having too much, not too little cryoprotectant agent in the cryo-buffer, as the crystal can then be subsequently annealed.

9. Conclusions on using cryo-crystallographic techniques

1. At 100 K, crystal lifetime is effectively infinite on a rotating anode and significantly extended at a synchrotron.

2. Better (i.e., more accurate and therefore more reliable) data can be collected with lower systematic errors:

    (a) Higher $I/\sigma(I)$ (longer counting times and higher signal-to-noise ratios)

    (b) Higher redundancy

    (c) Less absorption

    (d) Stronger anomalous data

    (e) Higher-resolution data can be collected (N.B. B values are reduced)

3. Very good for plate-like crystals and thin crystals: no stress

4. Can store crystals in liquid nitrogen at 78 K and collect more data later.

5. Need to invest time and crystals to determine the optimum cooling conditions. In many cases, this can be done on a rotating anode. It is very unwise to go to a synchrotron unprepared. The minimum cryoprotection (for glycerol, ethylene glycol, PEG400, and 1,2-propanediol concentrations for the "Magic 50" crystallisation solutions I and II can be found in [46, 47].

6. The mosaic spread of the crystal often increases to an unacceptable level, but in general room temperature mosaicity can be obtained by careful tuning of the cryoprotectant conditions [31].

7. "Small" crystals (surface to volume ratio of more than 12 mm^{-1}) are generally better, giving lower mosaic spread. They are more mechanically robust, and cool faster due to their larger surface to volume ratio.

8. The volume of the unit cell often changes on cooling (e.g., 2.4% decrease for glycogen phosphorylase b). Thus, non-isomorphism of heavy atom derivatives is still a major problem, but perhaps no worse than at room temperature (circumstantial comment from my particular experience, which has not been rigorously tested in general).

9. Need to minimise ice formation: take care about draughts and turbulence; use a coaxial room temperature dry nitrogen or dry air stream on the 100 K nitrogen stream.

10. Data for the multi-wavelength anomalous dispersion (MAD) method of phase determination can all be collected from the same crystal. This gives a much higher success rate.

11. It is hoped that more systematic approaches to treating problem crystals will develop, as well as automation of cryo-crystallographic techniques.

Cryocrystallography is also of value for investigating reaction rates of macromolecules with various compounds, as well as for substrate binding studies, since the reaction rates usually slow down significantly at low temperature, and data can be collected on crystals and substrates which would turn over too fast at room temperature. Crystals can also be cooled at various times after the start of, for example, reduction, to map intermediate conformational changes.

Flash-cooling is a now a *vital* technique for protein crystallography, since some projects are now feasible which were not possible at room temperature. MAD is now more likely to succeed, and significantly higher resolution data are obtainable, leading to higher resolution structures which in turn give more detailed and thus better, biological information.

References

1. Garman, E.F. and Schneider, T.R. (1997) *Journal of Applied Crystallography*, **27**: 211–237.
2. Rodgers, D.W. (1997) *Methods in Enzymology*, **276**: 183–202.
3. Garman, E. (1999) *Acta Crystallographica*, **D55**: 1641–1653.
4. Rodgers, D. (2001) In *International Tables for Crystallography: Volume F, Crystallography of Biological Macromolecules*. Edited by Rossmann, M.G. and Arnold, E., vol. F. Dordrecht: Kluwer Academic, pp. 202–208.

5. Garman, E.F. and Doublie, S. (2003) *Methods Enzymol*, **368**: 188–216.
6. Garman, E. and Owen, R.L. (2006) *Acta Crystallographica*, **D62**: 32–47.
7. Jones, G.D., Lea, J.S., Symons, M.C., and Taiwo, F.A. (1987) *Nature*, **330**: 772–773.
8. Weik, M., Ravelli, R.B., Kryger, G., McSweeney, S., Raves, M.L., Harel, M., Gros, P., Silman, I., Kroon, J., and Sussman, J.L. (2000) *Proceedings of the National Academy of Sciences of the USA*, **97**: 623–628.
9. Ravelli, R.B. and McSweeney, S.M. (2000) *Structure Folding and Design* **8**: 315–328.
10. Burmeister, W.P. (2000) *Acta Crystallographica*, **D56**: 328–341.
11. Garman, E. (2003) *Current Opinion in Structural Biology*, **13**: 545–551.
12. Nave, C. and Garman, E.F. (2005) *Journal of Synchrotron Radiation*, **12**: 257–260.
13. Hope, H. (1988) *Acta Crystallographica*, **B44**: 22–26.
14. Teng, T.-Y. (1990) *Journal of Applied Crystallography*, **23**: 387–391.
15. King, M.V. (1958) *Nature (London)*, **181**: 263–264.
16. Low, B.W., Chen, C.C.H., Berger, J.E., Singman, L., and Pletcher, J.F. (1966) *Proceedings of the National Academy of Sciences of the USA*, **56**: 1746–1750.
17. Haas, D. (1968) *Acta Crystallographica*, **B24**: 604–605.
18. Haas, D. and Rossmann, M.G. (1970) *Acta Crystallographica*, **B26**: 998–1004.
19. Thomanek, U. et al. (1973) *Acta Crystallographica*, **A29**: 263–265.
20. Kim, C.U., Kapfer, R., and Gruner, S.M. (2005) *Acta Crystallographica*, **D61**: 881–890.
21. Petsko, G.A. (1975) *Journal of Molecular Biology*, **96**: 381–392.
22. Teeter, M.M. and Hope, H. (1986) *Annals of the New York Academy of Sciences*, **482**: 163–165.
23. Parkin, S. and Hope, H. (1998) *Journal of Applied Crystallography*, **31**: 945–953.
24. Skrzypczak-Jankun, E., Bianchet, M.A., Amzel, L.M., and Funk, M.O. Jr. (1996) *Acta Crystallographica*, **D52**: 959–965.
25. Nagata, C., Other1, and Other2 (1996) *Acta Crystallographica*, **D52**: 623–630.
26. Fernandez, E.J. et al. (2000) *Journal of Applied Crystallography*, **33**: 168–171.
27. Garman, E.F. (1999) In *Protein Crystallisation: Techniques, Strategies and Tips*. Edited by Bergfors, T.M. La Jolla, CA: International University Line.
28. Thorne, R.E., Stum, J., Kmetko, K., O'Neill, R., and Gillilan, J. (2003) *Journal of Applied Crystallography*, **36**: 1455–1460.
29. Gregoriou, M., Noble, M.E., Watson, K.A., Garman, E.F., Krulle, T.M., de la Fuente, C., Fleet, G.W., Oikonomakos, N.G., and Johnson, L.N. (1998) *Protein Science*, **7**: 915–927.
30. Crick, F.H.C. and Magdoff, B.S. (1956) *Acta Crystallographica*, **D9**: 901–908.
31. Mitchell, E.M. and Garman, E.F. (1994) *Journal of Applied Crystallography*, **27**: 1070–1074.
32. Kwong, P.D. and Lui, Y. (1999) *Journal of Applied Crystallography*, **32**: 102–105.
33. Riboldi-Tunnicliffe, A. and Hilgenfeld, R. (1999) *Journal of Applied Crystallography*, **32**: 1003–1005.
34. Lusty, C.J. (1999) *Journal of Applied Crystallography*, **32**: 106–112.
35. Wierenga, R.K., Zeelan, J.P., and Noble, M.E. (1992) *Journal of Crystal Growth*, **122**: 231–234.
36. Ray, W.J. Jr., Baranidharan, S., and Liu, J. (1997) *Acta Crystallographica*, **D53**: 385–391.
37. Teng, T.-Y. and Moffat, K. (1998) *Journal of Applied Crystallography*, **31**: 252–257.
38. Walker, L.J., Moreno, P.O., and Hope, H. (1998) *Journal of Applied Crystallography*, **31**: 954.
39. Kriminski, S., Kazmierczak, M., and Thorne, R.E. (2003) *Acta Crystallographica*, **D59**: 697–708.
40. CRC (1988–1989) *Chemical Rubber Company Handbook*, 69th edn. Table D-232.
41. Yeh, J.I. and Hol, W.G.J. (1998) *Acta Crystallographica*, **D54**: 479–480.
42. Harp, J.M., Timm, D.E., and Bunick, G.J. (1998) *Acta Crystallographica*, **B54**: 622.
43. Harp, J.M., Hanson, B.L., Timm, D.E., and Bunick, G.J. (1999) *Acta Crystallographica Section D: Biological Crystallography*, **55**: 1329.

44. Kriminski, S., Caylor, C.L., Nonato, M.C., Finkelstein, K.D., and Thorne, R.E. (2002) *Acta Crystallographica*, **D58**: 459–471.
45. Juers, D.H. and Mathews, B.W. (2004) *Acta Crystallographica*, **B60**: 412–421.
46. Garman, E.F. and Mitchell, E.M. (1996) *Journal of Applied Crystallography*, 29: 584–587.
47. McFerrin, M. and Snell, E. (2002) *Journal of Applied Crystallography*, 35: 538–545.

PROCESSING DIFFRACTION DATA WITH MOSFLM

ANDREW G.W. LESLIE AND HAROLD R. POWELL

*MRC Laboratory of Molecular Biology, Hills Road,
Cambridge CB2 0QH, UK*

Abstract: Processing diffraction data falls naturally into three distinct
steps: First, determining an initial estimate of the unit cell and orientation
of the crystal; second, obtaining refined values for these parameters; and
third, integrating the diffraction images. The basic principles underlying
autoindexing, parameter refinement, and spot integration by summation
integration and profile fitting are described.

Keywords: data processing; profile fitting; autoindexing; postrefinement.

1. Introduction

This chapter will describe in outline the procedure for integrating
monochromatic diffraction data from macromolecules. It is assumed that the
diffraction images have been collected using the rotation method. Although the
procedures will be described with reference to the MOSFLM program, the basic
principles involved are common to most, if not all, data integration programs
currently in use. More detailed accounts of many aspects of data processing are
covered in the proceedings of a recent CCP4 Study Weekend [1].

2. Collecting the images

While the focus of this chapter is on data integration rather than data
collection, it is worth emphasizing that successful data integration depends
on the choice of appropriate experimental parameters during data collection. It is therefore crucial that the diffraction experiment is correctly
designed and executed. A list of the most important issues that need to be
considered is given below.

- Is the crystal single? Is the diffraction highly anisotropic? Two diffraction
  images 90° apart in phi should be examined carefully for evidence of split

41

R.J. Read and J.L. Sussman (eds.): Evolving Methods for Macromolecular Crystallography, 41–51
© 2007. *Springer*

spots or the presence of a second lattice. A single image can easily be misleading in this respect.

- Can the image be successfully indexed? Failure of the indexing could indicate the presence of a second lattice. Does the derived cell and orientation account for all the spots on the image (with an appropriate mosaic spread)? Are there lines of weak spots between those predicted (indicative of a pseudocell)? Are there additional spots due to the presence of a satellite crystal?

- Does the crystal really diffract to the edge of the detector? If not, either increase the exposure time or move the detector further away to improve the data quality (signal to noise).

- Is the collimation adequate to resolve adjacent reciprocal lattice spots for the longest cell spacing? If not, move the detector further back or try reducing the incident beam size or, in some circumstances, the beam divergence.

- Is the dynamic range of the detector sufficient to avoid overloaded reflections at low resolution? If not, a rapid pass may be necessary to measure these strong reflections. Ideally, collect this rapid pass first.

- What is the optimum rotation angle per image? Too large a value will result in spatial overlap of spots in adjacent lunes. Too small a value will give a poor duty cycle, as the exposure time becomes comparable with the detector readout time. Very short exposure times (less than ~0.5 s) on modern synchrotron sources can lead to problems with shutter synchronization.

- Ideally, aim for high data multiplicity as this will improve the overall quality of the data by reducing random errors and facilitating outlier identification. If this is not possible, aim for high completeness, possibly by collecting several segments of data rather than a single large rotation. Be conservative in the choice of exposure time, so that the data set is complete before the onset of serious radiation damage.

- Always integrate at least some (and preferably all) the diffraction images during data collection, to check for unforeseen problems and to get a quantitative estimate of data quality. Soon it should be possible to do this automatically.

3. Determining the crystal cell parameters and orientation

The autoindexing algorithms currently in use are extremely powerful and in general it will be possible to determine the unit cell dimensions and crystal orientation from a single diffraction image, providing that the direct beam position, crystal to detector distance, and radiation wavelength are accurately known. Failure of the autoindexing can result from errors in these experimental parameters, the presence of a second lattice, or if only very few spots are available for the autoindexing. In the last case, inclusion of spots from two or more images should lead to success.

The most robust autoindexing algorithms employ a Fourier transform approach [2]. The general principle behind Fourier-based autoindexing can be understood as follows. Figure 1 indicates a situation where a "still" image (i.e., zero oscillation angle) has been taken with the crystal in an orientation such that a principle zone axis lies along the X-ray beam direction. The planes of reciprocal lattice points normal to this zone axis intersect the Ewald sphere in a series of concentric circles, centered on the direct beam position. In the diffraction image, a series of concentric lunes will be seen. Using the Ewald sphere construction, all of the spots on the detector can be mapped back to the reciprocal lattice points that gave rise

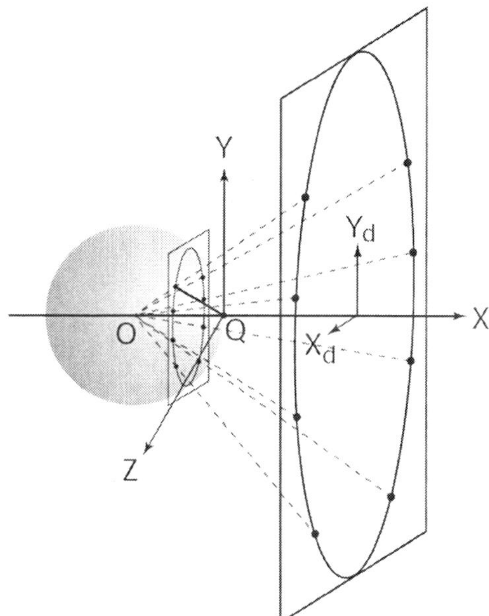

Figure 1. The Ewald sphere construction showing the geometry of diffraction for the case in which a principle zone axis lies along the X-ray beam direction. X, Y, Z define the laboratory coordinate frame. X-rays are parallel to the X-axis, and the phi rotation axis is along Z. The crystal sits at position O. A single plane of the reciprocal lattice normal to the X-ray beam is shown. Within this plane, reciprocal lattice points that lie on the Ewald sphere and are therefore in a diffracting condition are shown as dots. The resulting spots on the detector face are also shown. Knowing the detector geometry and the spot coordinates, the positions of the reciprocal lattice points can be calculated. A representative scattering vector (from the origin of the reciprocal lattice (marked Q) to the reciprocal lattice point) is shown.

to those spots (there is a small error involved in doing this, as the actual phi value for each reflection is not known). Now consider what happens when all of these "scattering vectors" are projected onto the zone axis. All the spots lying within the same lune will give rise to a projected vector of the same length. Thus, the projected scattering vectors for all the spots on the image will fall into clusters, where the separation between each cluster corresponds to the vector between adjacent reciprocal lattice planes. The Fourier transform of the projected clusters will form a series of regularly spaced spikes (Figure 2), where the distance between adjacent spikes corresponds to the real cell spacing along the principal zone axis direction. Now consider projecting the scattering vectors along a direction at an angle of (say) 20° to the true zone axis direction. In this case, spots in the same lune will project to give vectors of *different* lengths and so the Fourier transform of the projected scattering vectors will *not* have the clear set of maxima shown in Figure 2.

In practice [2, 3], the direction of the projection axis is varied in small angular steps (e.g., 2°) for the complete hemisphere of directions and in each case the Fourier transform of the projected scattering vectors is calculated.

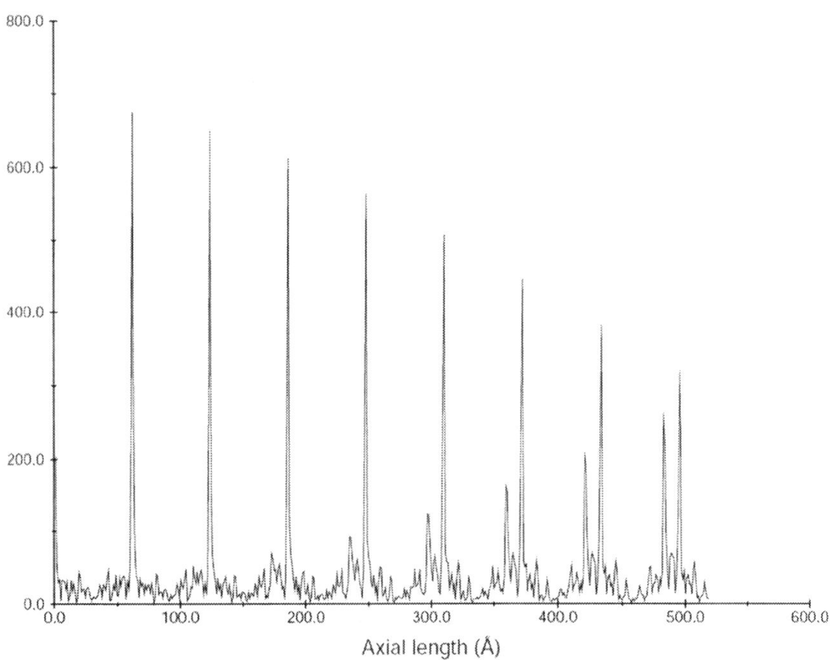

Figure 2. The Fourier transform of the projected scattering vectors for the case shown in Figure 1 will consist of a number of regularly spaced discrete maxima, where the spacing between adjacent peaks reflects the real cell spacing along the zone axis direction.

Then three directions are chosen from this list that have large maxima in the Fourier transform and reasonably large interaxial angles. These will define three principle zone axes and their repeats, thus defining a unit cell with which it should be possible to index all spots in the diffraction image. In general, the resulting unit cell will be a triclinic one that will not reflect the true symmetry of the lattice. The final stage is therefore to find the reduced cell from the chosen cell and then evaluate a "goodness of fit" to the 44 possible lattice types [4, 5]. The user is presented with a list of possible solutions, each with a corresponding quality index and, in general, the solution with the highest Bravais lattice symmetry that still has a good quality index will be chosen. It is important to realize that there is no information available at this stage on the *true* crystal symmetry, which can only be determined from the diffraction intensities. The spot positions only give information about the lattice symmetry, which can be higher than the true crystal symmetry. This is particularly important when considering the strategy for data collection. An incorrect assumption about the crystal symmetry may lead to the choice of a total rotation angle that is too small to collect all the unique data. For example, there are numerous examples of monoclinic crystals with a β angle very close to 90°. If the symmetry is incorrectly assumed to be orthorhombic and only 90° of data are collected rotating around the *b*-axis, then the resulting data will be very incomplete.

4. Parameter refinement

Once an orientation matrix and cell parameters have been derived from the autoindexing, these parameters (and others) are refined further using different algorithms. The parameters to be refined can be conveniently grouped into three classes:

- Crystal parameters: cell parameters, crystal orientation, and mosaic spread (isotropic or anisotropic)
- Detector parameters: the detector position and orientation and (if appropriate) distortion parameters (e.g., the radial and tangential offsets for the Mar image plate scanner)
- Beam parameters: the orientation of the primary beam and beam divergence (isotropic or anisotropic)

There are two complementary sources of information that can be used in the refinement; the spot coordinates measured on the detector, and the spot coordinates in phi. The latter can be measured empirically if the oscillation angle is much smaller than the reflection width, or can be estimated from the way in which the intensity for partially recorded reflections is distributed over the two (or more) images on which the reflection is recorded if the oscillation angle is comparable to, or greater than, the reflection width.

4.1. REFINEMENT USING SPOT COORDINATES MEASURED ON THE DETECTOR

The parameters are refined by least squares minimization of a positional residual:

$$\Omega_1 = \sum_i w_{ix}\left(X_i^{\text{calc}} - X_i^{\text{obs}}\right)^2 + w_{iy}\left(Y_i^{\text{calc}} - Y_i^{\text{obs}}\right)^2 \tag{1}$$

where X and Y are the spot coordinates on the detector, and w_{ix} and w_{iy} are appropriate weights.

Note that it is not possible to refine changes in crystal orientation around the rotation axis using this residual, as this parameter has no effect on the spot positions. Other parameters, such as cell dimensions and crystal to detector distance, may also be highly correlated (depending on the maximum Bragg angle).

4.2. REFINEMENT USING PHI COORDINATES

In this case, the residual to be minimized is given by:

$$\Omega_2 = \sum_i w_i\left[\left(R_i^{\text{calc}} - R_i^{\text{obs}}\right)/d_i^*\right]2 \tag{2}$$

where R_i^{calc} and R_i^{obs} are the calculated and observed distances of the reciprocal lattice point d_i^* from the center of the Ewald sphere (OP and OP′ in Figure 3) and again w_i is a weighting term. R_i^{calc} is determined from the current values for the cell parameters and crystal orientation. R_i^{obs} is obtained from the Φ centroid if fine Φ slices have been used. For coarse Φ slices, the position in phi of partially recorded reflections is estimated from the degree of partiality of the reflection (i.e., the way in which the total intensity is distributed between the two (or more) abutting images). This latter approach, known as postrefinement [6, 7] because it depends on knowing the integrated intensities, requires a model for the rocking curve, and permits refinement of either crystal mosaicity or beam divergence.

The effective radius of the reciprocal lattice point (see Figure 3) is given by

$$\varepsilon = \frac{\gamma d^*}{2}\cos\theta \tag{3}$$

where γ is the combined mosaic spread and beam divergence, d^* is the reciprocal lattice spacing and θ is the Bragg angle. The distance of the reciprocal lattice point from the Ewald sphere, Δr, is then given by

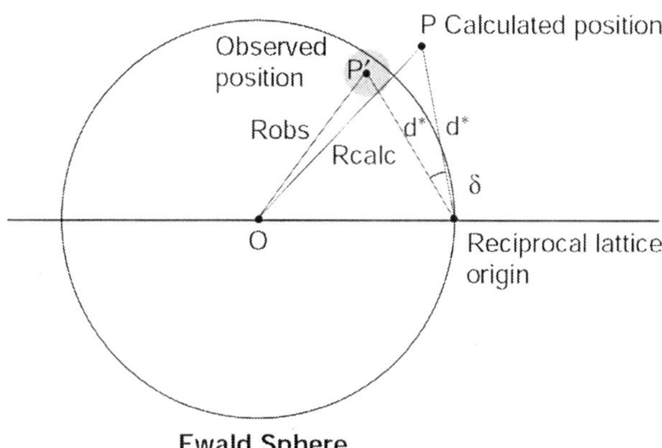

Figure 3. The large circle represents a section through the Ewald sphere, while the small shaded circle represents the position of a reciprocal lattice point at the end of an oscillation. A fraction of the total intensity corresponding to the volume of the reciprocal lattice point that has already passed through the Ewald sphere will be recorded on the current image. The remaining intensity will be recorded on the following image as P′ rotates clockwise.

$$P = 1/2 \left[1 + \sin \left(\pi \Delta r / 2\varepsilon \right) \right] \tag{4}$$

$$\text{where } P = \frac{I_1}{I_1 + I_2} \tag{5}$$

and I_1, I_2 are the intensities recorded on the two abutting images (assuming the reflection only spans two images). Knowing P from the measured intensities, Δr can be calculated from equation 4, and thus R^{obs} can be determined. Rocking curve models other than the simple sine model in equation 4 have also been used. Because ε depends on the combined mosaic spread and beam divergence, this parameter can also be refined. (For fine Φ slices the mosaic spread or beam divergence is estimated from the observed reflection width in Φ.)

4.3. REFINEMENT STRATEGY

The refinement strategy can depend on how the data has been collected. If fine Φ slices have been used, accurate Φ centroids and coordinates (X, Y) are available for most strong reflections (excluding those very close to the rotation axis) and both residuals (Ω_1, Ω_2) can be minimized simultaneously using a suitable selection of reflections (strong and evenly distributed over the detector and in Φ). Problems arising due to correlations of different parameters can be avoided either by fixing some parameters or by the use of eigenvalue filtering. These problems can be particularly serious for low resolution data, where there is a strong correlation between crystal to detector distance and the cell parameters, or for an offset detector where there is a high correlation between the detector swing angle and the (horizontal) primary beam coordinate. If only a narrow Φ range of reflections is used in the refinement then some unit cell parameters will be poorly defined and may be correlated with the crystal setting angles, and there will also be a strong correlation between the detector orientation around the X-ray beam and the crystal setting angle around the beam. In such circumstances, the refined parameters may assume physically unrealistic values, but this will not necessarily affect the accuracy of the prediction of reflection positions and widths.

When the data is collected with coarse Φ slices, only fully recorded reflections will give accurate spot positions (X, Y), and accurate Φ centroids can only be determined for partially recorded reflections. In MOSFLM, the two residuals are currently minimized independently. Only the detector parameters are refined when minimizing the positional residual, and only cell, orientation and optionally beam parameters are refined against the angular residual. This approach does have the advantage that the accuracy of the refined cell parameters does not depend on the accuracy of the crystal to detector distance or direct beam position, providing these are known sufficiently well to allow correct indexing of the reflections.

5. Integration of the images

Once accurate values for the crystal cell parameters and orientation have been obtained, the images can be integrated. Stated in the simplest way, this procedure involves predicting the position in the digitized image of each Bragg reflection present on that image, and then estimating its intensity (after subtracting the X-ray background) and an error estimate of the intensity. In practice, this apparently simple task is quite complex.

5.1. PREDICTING REFLECTION POSITIONS

A knowledge of the crystal cell and orientation will allow the prediction of spot positions on a "virtual detector," i.e., a detector whose position and orientation are exactly known. These positions must then be mapped onto the digitized image, and this mapping must take into account any spatial distortions introduced by the detector, either using a predetermined calibration table or by refining the distortion parameters for each image.

5.2. DEFINING THE PEAK/BACKGROUND MASK

Because it is physically impossible to measure the X-ray background actually under the diffraction spot (which strictly is what is required to obtain the background subtracted intensity) the background is measured in a region around the spot either in two dimensions (X, Y, the detector coordinates) for coarse Φ slices or in three dimensions (X, Y, and Φ) for fine Φ slices. A background plane is fitted to these background pixels, and this plane is then used to estimate the background under the spot. To do this it is necessary to define a pixel mask, which, when centered on the predicted position of the spot, will define which pixels are to be considered as part of the peak and which are to be used to determine the background (see Figure 4).

This mask can be defined by the user after visual inspection of the spot shapes, but MOSFLM will automatically optimize the peak/background definition. It is clearly important that pixels are not misclassified, as this can lead to systematic errors in the integrated intensity. The presence of strong diffuse scattering, which is quite commonly observed with synchrotron data, can lead to difficulties in differentiating between peak and background pixels. Unfortunately, there is no simple way of dealing with this problem.

5.3. SUMMATION INTEGRATION AND PROFILE FITTING

Having determined the background plane, the simplest way to obtain an estimate of the integrated intensity is to sum the pixel values of all pixels in the peak area of the mask, and then subtract the sum of the background values calculated from the background plane for the same pixels. This is known as summation integration and for spots where the background level is very low

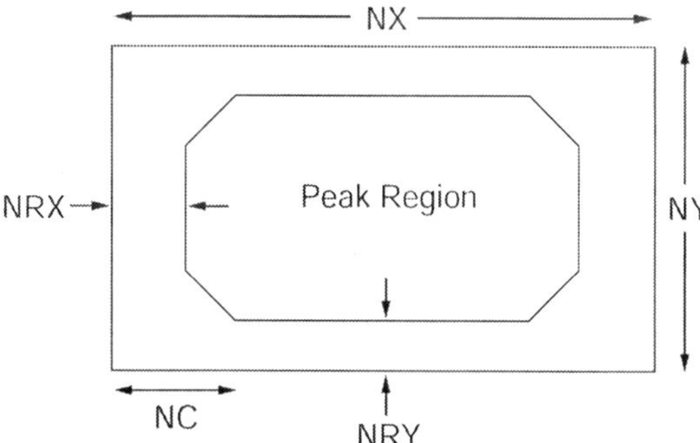

Figure 4. The peak/background mask definition used in MOSFLM. The overall mask size (in pixels) is defined by NX and NY, and the differentiation between peak and background pixels is defined by a background rim in X and Y (NRX, NRY pixels) and a corner cutoff (NC pixels).

compared to the intensity of the spot this will give as accurate an estimate of the intensity as it is possible to get. (In such cases, the accuracy is determined by counting statistics, so for a total count of N photons the standard deviation is $\sqrt{N}$).

For weaker reflections, it is possible to get a more accurate estimate of the integrated intensity by using a procedure known as profile fitting [8–11]. In this procedure, it is assumed that the shape or profile (in two or three dimensions) of the spots is known. The background plane is determined in the same way as for summation integration, but the intensity is derived by determining the scale factor which, when applied to the *known* spot profile, gives the best fit to the *observed* spot profile. This scale factor is then proportional to the profile fitted intensity for the reflection. In practice, the fitting is done by least squares methods, to minimize the residual

$$R = \sum_{\substack{\text{peak} \\ \text{pixels}}} w_i \left(X_i - KP_i \right)^2 \qquad (6)$$

where

     X_i is the background subtracted intensity at pixel i

     P_i is the value of the standard profile at the corresponding pixel

     w_i is a weight, derived from the expected variance of X_i

     K is the scale factor to be determined.

The improvement gained by profile fitting depends on the spot intensity relative to background and the spot shape, but typically it can provide a reduction in variance by a factor of 2 (1.4 in the standard deviation) for weak reflections. This is a significant gain, and all modern software packages employ profile fitting, although the implementation differs in detail.

The procedure assumes that the *true* reflection profile is known. In practice, this is determined from the observed reflection profiles of a number of reflections in the immediate vicinity of the reflection being integrated. An appropriate weighted sum of the individual profiles is used to form the "true" or standard profile. The reflection shape will vary with position on the detector (due to changes in obliquity of incidence and other factors) and it is important to allow for this. MOSFLM determines a "standard" profile for several defined areas and then calculate the best profile for each reflection as a weighted mean of the closest "standard" profiles.

Profile fitting is a powerful technique for reducing the random error in weak diffraction data, but equally an error in determining the standard profiles will lead to systematic errors in all measured intensities. Modern software packages go to some lengths to minimize the magnitude of the systematic errors introduced by the use of nonideal standard profiles.

5.4. STANDARD DEVIATION ESTIMATES

It is important to obtain reasonable estimates of the standard deviations of the integrated intensities, since these are used as weights when merging multiple observations, and in subsequent steps of the structure determination (e.g., identification of heavy atom derivatives, heavy atom parameter refinement, and model refinement). For summation integration, a standard deviation can be obtained based on Poisson statistics, while for profile fitted intensities the goodness of fit of the scaled standard profile to the true reflection profile can be used. These will generally underestimate the true errors, as they take no account of systematic errors arising from effects such as absorption, beam instability, detector nonlinearity, or errors in nonuniformity corrections. The standard deviation estimates should therefore be modified when the data is merged, making use of the *observed* agreement between multiple observations.

References

1. Multiple contributions (1999) *Acta Crystallographica*, **D10**: 1631–1772.
2. Steller, I. et al. (1997) An algorithm for automatic indexing of oscillation images using Fourier analysis. *Journal of Applied Crystallography*, **30**: 1036–1040.
3. Powell, H.R. (1999) The Rossmann Fourier autoindexing algorithm in MOSFLM. *Acta Crystallographica*, **D10**: 1690–1695.
4. Burzlaff, H. et al. (1992) *International Tables for Crystallography*, vol. A. Edited by T. Hahn. Dordrecht: Kluwer Academic, pp. 737–749.
5. Kabsch, W. (1993) Automatic processing of rotation diffraction data from crystals of initially unknown symmetry and cell constants. *Journal of Applied Crystallography*, **26**: 795–800.
6. Winkler, F.K. et al. (1979) The oscillation method for crystals with very large unit cells. *Acta Crystallographica*, **A35**: 901–911.
7. Rossmann, M.G. et al. (1979) Processing and post-refinement of oscillation camera data. *Journal of Applied Crystallography*, **12**: 570–581.
8. Diamond, R. (1969) Profile analysis in single crystal diffractometry. *Acta Crystallographica*, **A25**: 43–54.
9. Ford, G.C. (1974) Intensity determination by profile fitting applied to precession photographs. *Journal of Applied Crystallography*, **7**: 555–564.
10. Rossmann, M.G. (1979) Processing oscillation diffraction data for very large unit cells with an automatic convolution technique and profile fitting. *Journal of Applied Crystallography*, **12**: 225–238.
11. Leslie, A.G.W. (1999) Integration of macromolecular diffraction data. *Acta Crystallographica*, **D10**: 1696–1702.

SAD PHASING: BASIC CONCEPTS AND HIGH-THROUGHPUT

GEORGE M. SHELDRICK

*Lehrstuhl für Strukturchemie, Universität Göttingen,
Tammannstraße 4, 37077 Göttingen, Germany*

Abstract: Various fundamental concepts involved in experimental phasing
are discussed with special attention to SAD phasing. They are illustrated
briefly by reference to the programs SHELXC, SHELXD, and SHELXE
that are often used in high-throughput structure solution pipelines.

Keywords: phasing; SAD; SHELX; Patterson methods; density modification.

1. Introduction

In keeping with the character of this Erice School, some concepts important
for the understanding of experimental phasing of macromolecules will be
discussed. Rather than describing again methods that have already been
well documented, we will concentrate on topics that are essential for a
good understanding of experimental phasing, but are easily forgotten in
the excitement of solving a new structure using the latest automated
pipelines.

Except in relatively rare cases, where atomic resolution data permit
the phase problem to be solved by *ab initio* direct methods, experimental
phasing usually implies the presence of *heavy atoms* to provide *reference
phases*. We then calculate the phase ϕ_T of each reflection for the full
structure by:

$$\phi_T = \phi_A + \alpha$$

Where ϕ_A is the calculated phase of the heavy-atom substructure for that
reflection. As we will see, α can be estimated from the experimental data.
Of course, it is better to use a probability distribution than a single phase, but
we will concentrate on individual phases because they are easier to understand and in any case are still required whenever we wish to calculate an
electron density map. The experimental phase determination requires the
following stages:

53

R.J. Read and J.L. Sussman (eds.): Evolving Methods for Macromolecular Crystallography, 53–65
© 2007. *Springer*

1. Location of the heavy atoms

2. Refinement of heavy atom parameters (if necessary) and calculation of ϕ_A

3. Calculation of starting protein phases using $\phi_T = \phi_A + \alpha$

4. Phase improvement by density modification (and where appropriate non-crystallographic symmetry [NCS] averaging).

2. SAD, SIR, and SIRAS in the light of MAD phasing

Karle [1] and Hendrickson et al. [2] showed by algebra that the measured intensities in a MAD experiment should be given by:

$$|F_+|^2 = |F_T|^2 + a|F_A|^2 + b|F_T||F_A|\cos\alpha + c|F_T||F_A|\sin\alpha$$

and

$$|F_-|^2 = |F_T|^2 + a|F_A|^2 + b|F_T||F_A|\cos\alpha - c|F_T||F_A|\sin\alpha$$

where $|F_+|^2$ is the measured intensity of reflection h,k,l and $|F_-|^2$ that of $-h,-k,-l$; $a = (f''^2 + f'^2)/f_0^2$, $b = 2f'/f_0$ and $c = 2f''/f_0$. a, b, and c are different for each wavelength. In a MAD experiment, provided that $|F_+|^2$ and $|F_-|^2$ have been measured at two or more wavelengths we can extract $|F_T|$ (native F including heavy atoms but ignoring f' and f'' contributions), $|F_A|$ (experimental heavy atom structure factor), and α (phase shift from heavy atom phase to protein phase) for each reflection. All we then need to do is to find the heavy atoms (from $|F_A|$), use them to get ϕ_A, then calculate a map with amplitudes $|F_T|$ and phases $\phi_T = \phi_A + \alpha$. For SIRAS, we can also deduce $|F_T|$, $|F_A|$ and α from similar equations, so the situation is similar to MAD, but because of possible lack of isomorphism the resulting protein phases are usually not quite so good.

 For SAD, by introducing the approximation $|F_T| \sim \tfrac{1}{2}(|F_+| + |F_-|)$ we can derive $|F_+| - |F_-| = c|F_A|\sin\alpha$, and for SIR we note that $|F_T| = |F_{nat}|$ and (assuming that the isomorphous differences are small compared with $|F_T|$) we obtain $|F_{deriv}| - |F_{nat}| = b|F_A|\cos\alpha$.

 Programs that find the heavy-atom substructure such as SHELXD have to use $c|F_A|\sin\alpha$ instead of $|F_A|$ as coefficients in the case of SAD. At first sight it is surprising that this is so effective at finding the sites. In using *direct methods* to find these sites the $|c|F_A|\sin\alpha|$ amplitudes are normalized to give E-values so the (resolution-dependent) constant c is eliminated; only the largest E-values are used. Statistically the E-values with $\sin\alpha$ close to ± 1 have larger absolute values and so are more likely to be selected, so the approximation is better than expected. Since the anomalous differences at high resolution can be drowned by the noise, it may be necessary to truncate the resolution for locating the heavy atoms (but not for the subsequent density modification).

When the heavy atoms have been found, in principle we could replace $|F_A|$ in the equation $|F_+|-|F_-| = c|F_A|\sin\alpha$ by the calculated heavy atom structure factor $|F_H|$ and we would be able to calculate $\sin\alpha$. This defines α except for a twofold ambiguity. In practice, experimental errors and difficulties in scaling and estimating c make this far from straightforward.

In SHELXE [3], all amplitudes are first normalized to give E-values. Large normalized anomalous differences must be associated with values of $\sin\alpha$ close to ± 1, and an empirical figure of merit (fom) in the range 0–1 is assigned to indicate how reliable these estimates of $\alpha = 90°$ (when $|F_+| \gg |F_-|$) or 270° (when $|F_-| \gg |F_+|$) are. Reflections that have a large normalized $|F_H|$ but a small normalized anomalous difference will have phases close to either 0° or 180°, but we cannot tell which. In most programs this is represented by a bimodal probability distribution (e.g., using Hendrickson–Lattman coefficients [4]). SHELXE makes an attempt to resolve these twofold ambiguities by a special density modification cycle in which all but the highest density values are set to zero before Fourier inversion, and uses this to assign initial phases in the full range 0–360° with appropriate figures of merit.

3. The heavy atom enantiomorph problem

The location of the heavy atoms from the $|F_A|$-values does not define the enantiomorph of the heavy-atom substructure; there is exactly a 50% chance of getting the enantiomorph right. In general, both the heavy-atom substructure and the inverted substructure need to be considered. If the space group is one of an enantiomorphic pair (e.g., $P4_12_12$ and $P4_32_12$) the space group must be inverted, as well as the atom coordinates. For three of the 65 space groups possible for chiral molecules, the coordinates have to be inverted in a point other than the origin. These space groups and inversion operations are:

$$I4_1\ (1-x, \tfrac{1}{2}-y, 1-z);\ I4_122\ (1-x, \tfrac{1}{2}-y, \tfrac{1}{4}-z);\ F4_132\ (\tfrac{1}{4}-x, \tfrac{1}{4}-y, \tfrac{1}{4}-z).$$

Why do we have to invert a substructure in the space group $I4_1$ in, for example, the point ($\tfrac{1}{2}$, $\tfrac{1}{4}$, $\tfrac{1}{2}$) rather than in the origin? This problem is a bit like having to change space groups on inverting $P4_1$ to $P4_3$, except that $I4_1$ possesses 4_1 and 4_3 axes and so is its own enantiomorph. Inversion of all the coordinates in the origin changes the 4_1 axis of the space group to a 4_3 axis and so violates the space group as defined by its symmetry operators. Inversion in ($\tfrac{1}{2}$, $\tfrac{1}{4}$, $\tfrac{1}{2}$) not only leaves the symmetry elements as they are; it also inverts the arrangement of the atoms. A helix around the 4_3 axis becomes a helix about a 4_1 axis.

However, the role of the substructure enantiomorph is different for the different phasing methods (SAD, SIR, etc.). To understand this, we will look at maps of a hypothetical bromobenzene structure phased by one heavy atom using perfect data (Figure 1). These can be generalized to the case of a centrosymmetric arrangement of heavy atoms. A perfect MAD or SIRAS experiment should give perfect phases. The quality of SIRAS and MAD maps is limited only by the quality of the data, which is determined by radiation damage (MAD) and lack of isomorphism (SIRAS), as well as by the strength of diffraction from the crystals. A centrosymmetric array of heavy atoms is fatal for SIR because there is no escape from the resulting centrosymmetric double image of the structure.

For SAD, the map phased using only $\alpha = 90$ or $270°$ does not show the anomalous atoms when their arrangement is centrosymmetric (they belong to both the positive and negative images and so cancel) and consists of positive and negative images of the rest of the structure. The heavy atoms can simply be added as a contribution with phase ϕ_A, i.e., orthogonal to the SAD phases. The negative image can in principle be removed by replacing negative density with zero, but where the negative and positive images overlap it is not possible to recover density in this way. Although the stronger the anomalous

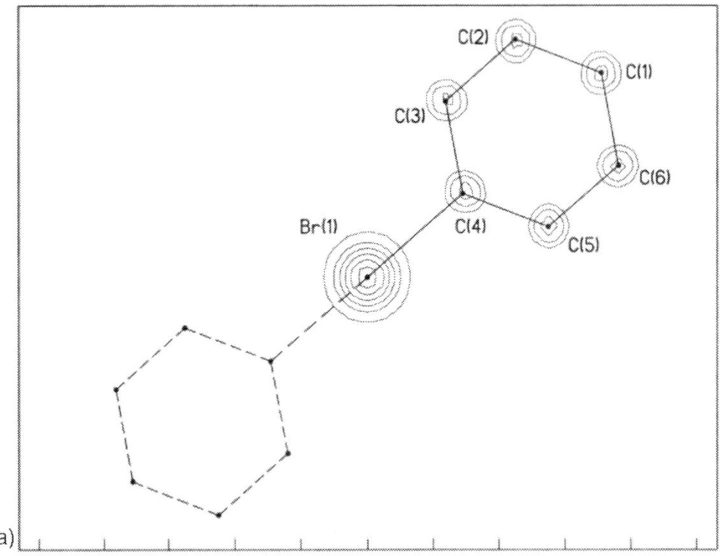

Figure 1. Phase determination using a single heavy atom in space group P1 and ideal data. (a) MAD or SIRAS phasing gives a perfect map.

(continued)

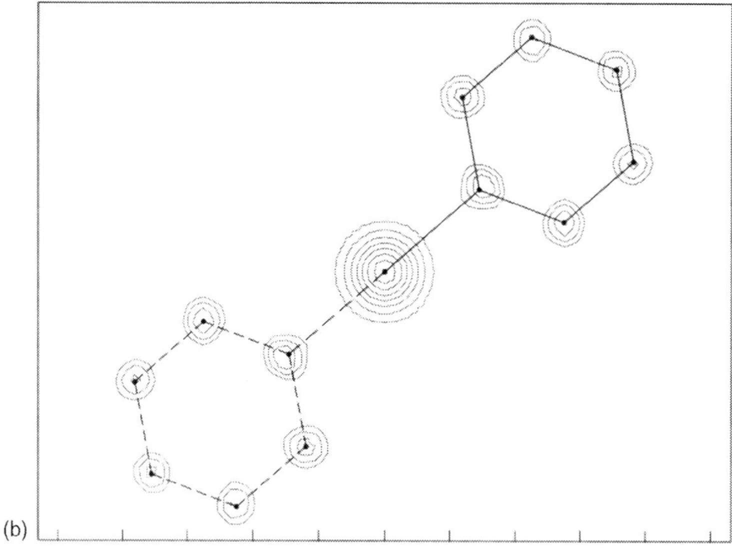

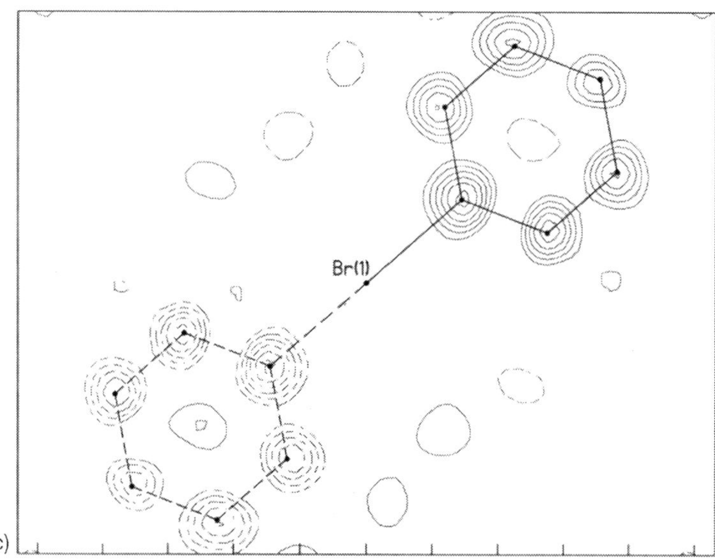

Figure 1. (continued) (b) SIR produces a centrosymmetric double image. (c) SAD leads to positive and negative images. The latter could easily be removed by density modification, but this would not reinstate density lost by overlap of the positive and negative images. This will be more of a problem if the solvent content and resolution are low.

signal, the better the map, even with perfect SAD data the map may not be interpretable without improvement by density modification.

With a chiral substructure, SIR gives the same figures of merit in the density modification for both heavy-atom enantiomers. In the best case, one map is correct and the other is its mirror image (the α-helices go the wrong way). With a centrosymmetric substructure, for SAD both heavy-atom enantiomorphs give the same (correct) structure and figures of merit. If the substructure is chiral, for SAD the figures of merit should identify the correct enantiomorph; one map should be correct and the other should be nonsense. Space groups that are members of enantiomorphous pairs (e.g., $P3_1$, $P4_12_12$) are particularly good for heavy atom location and density modification (strong enantiomorph definition). Even one-site SIR works well for such space groups.

4. Locating the heavy atoms

The same methods used for *ab initio* all atom structure solution from very high resolution native data turn out to be eminently suitable for the location of heavy atom sites from SIR, SAD ΔF, or MAD F_A values. The iterative dual-space method introduced in the Program SnB [5] and now used in SHELXD [6] and HySS [7] has been discussed in some detail [8] and so will not be reviewed again here. An important addition is the refinement of the occupancies of the sites in the final cycles, which allows both for varying occupancies (e.g., for iodide soaks) and for variations in B-values when phasing with sulfur and selenium atoms. The occupancy refinement is started with a few more sites than expected, so that a sharp falloff in the refined occupancy gives a good indication of the true number of sites. It suffices when the number of sites input is within about 20% of the true value; the SHELXD job should be repeated with a different number of expected sites if the occupancy refinement indicates that this is necessary.

In the case of sulfur-SAD, the two sulfurs in a disulfide bridge fuse into a single super-sulfur atom at resolutions worse than about 2.2 Å. However, the resolution to which the anomalous difference data are truncated (because these differences may be almost pure noise at the highest resolution attained) can be critical. In the case of weak SAD data, the optimum point to truncate will tend to be at about $d_{min} + 0.5$ Å, but some experimentation may be required to find the optimum value. If data have been measured from two crystals, a good point is where the correlation coefficient (CC) between the signed anomalous differences falls to below 30%; a similar test is often used for the data at two different wavelengths in a MAD experiment [9]. If data are only collected from one crystal, it is possible to split them randomly into two sets and calculate the CC between them; this has been implemented in SCALA and SHELXC, but the CCs may be overestimated because the data are less independent. The ratio of $||F_+|-|F_-||$ to its estimated standard deviation (esd) is a less reliable guide because it is difficult to obtain sufficiently accurate esds for the measured intensities.

Although the Patterson-seeded dual-space recycling in SHELXD is very effective and robust at finding the heavy-atom substructure, some thought should be given to the minimum distance to allow between two sites and as to whether sites should be allowed on special positions or not. Incorrect direct methods pseudosolutions often have peaks on special positions, but it is not unusual for iodide ions to be found on special positions in a soak, and it is even possible that an unanticipated zinc or other ion lies on a twofold axis. In difficult cases, it may be necessary to run more trials (say 5,000 rather than 100).

5. Probabilistic Patterson sampling

Each non-Harker Patterson vector of suitable length is a potential heavy atom to heavy atom vector, and may be employed as a two-atom search fragment in a translational search based on the *Patterson minimum function*. For each position of the two atoms in the cell, the Patterson height P_j is found for all vectors between them and their symmetry equivalents, and the sum (PSUM) of the lowest (say) 35% of P_j calculated. It would be easy to find the global maximum of PSUM using a fine three-dimensional (3D) grid, and this is an efficient way of solving small substructures. For large substructures, especially when pseudosymmetry is present, there may be severe overlap of the Patterson peaks, and simply taking the highest maxima of PSUM may well *not* lead to the solution of the structure. A more effective approach is to generate many different starting positions by simply taking the best of a finite number of random trials each time. This sampling has the advantage that it may be run as often as necessary, and will each time produce a new starting position for dual-space iteration that is more likely (in practice, by about one order of magnitude) to lead to a successful substructure solution than starting from random atom positions or phases. The *full-symmetry Patterson superposition minimum function* is used to expand from the two atoms to a much larger number before entering the dual-space recycling.

Patterson seeding tends to be less effective for space group P1, possibly because the dual-space iteration tends to have a high success ratio anyway in this space group, and in high symmetry (e.g., cubic) space groups, because the CPU time required becomes significant.

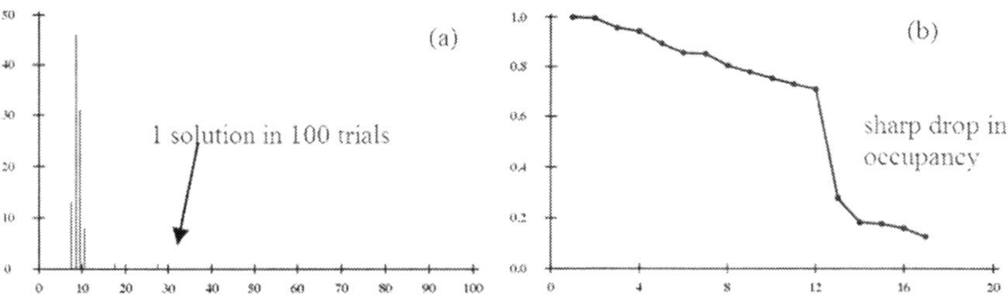

Figure 2. (a) Number of solutions against CC; (b) occupancy against site number. The sharp drop is indicative of a good solution despite the modest CC (27%). These diagrams are part of the standard output of the HKL2MAP GUI [13].

6.   An example of sulfur-SAD substructure determination

Crystals of elastase diffract well and are suitable for in-house sulfur-SAD structure solution. The distribution of CCs shows one solution with a CC of about 27 in 100 tries (Figure 2); it would be safer to increase the number of sets of random starting vectors in this case. The refined occupancies show a sharp drop after 12 sites (four disulfide bonds, two methionines, and two sulfate anions).

7.   The sphere of influence algorithm

SHELXE employs a novel algorithm to decide how likely a pixel in a map is an atomic site of the protein (or DNA, etc.) and how likely it is not (i.e., in the solvent region or at high resolution between the atoms).

The variance V of the density on a spherical surface of radius 2.42 Å is calculated for each pixel in the map. The use of a spherical surface rather than a spherical volume was intended to add a little chemical information (2.42 Å is a typical 1,3-distance in proteins and DNA). The pixels with the highest V are most likely to correspond to real protein atomic positions. Pixels with low V are *flipped* ($\rho' = -\rho\gamma$ where γ is about one). For pixels with high V, ρ is replaced by $\rho' = [\rho^4/(v^2\sigma^2(\rho) + \rho^2)]^{1/2}$ (with v usually 1) if it is positive and by zero if it is negative. This has a similar effect to the procedure used in the program ACORN [14]. A *fuzzy boundary* is used between these two regions; in the *fuzzy region* ρ' is set to a weighted sum of the two treatments. The *fuzzy boundary* is an attempt to allow the pixel assignments to change flexibly as the density modification proceeds. An empirical weighting scheme for phase recombination is used to combat model bias. Usually 15–20 cycles of this density modification reach a steady state, but in cases where high-resolution native data are phased starting from very low-resolution phases 100 or 200 cycles may be required. An example of the SHELXE density modification is shown in Figures 3 and 4.

The variance averaged over all pixels of V (referred to by SHELXE as *contrast*) should be larger for the correct heavy-atom enantiomorph and also indicates when the density modification has converged. A related criterion is used by Terwilliger [15] in the program SOLVE.

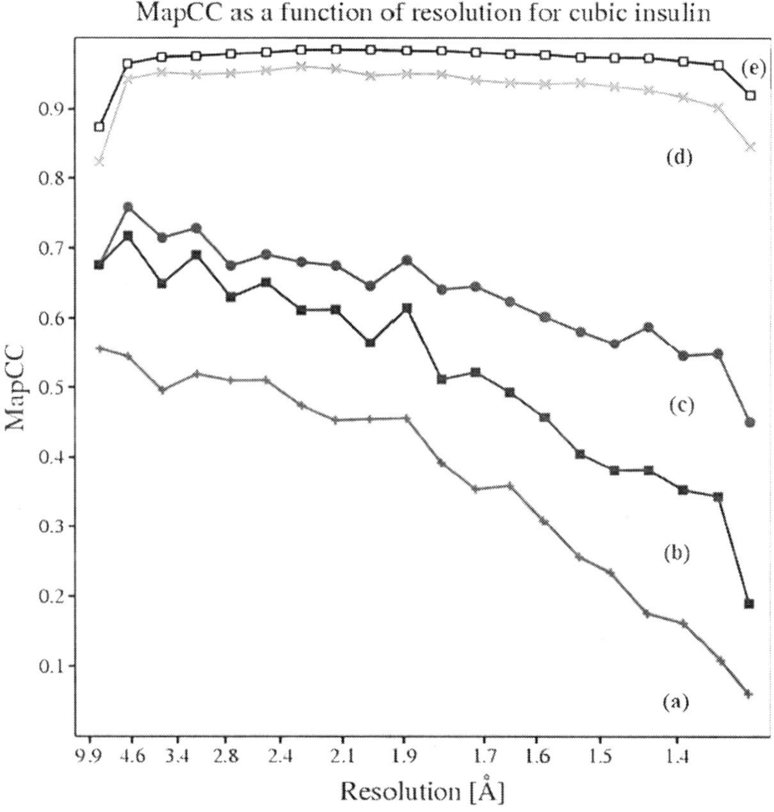

Figure 3. Stages in the SHELXE phasing and density modification of cubic insulin. (a) Map correlation coefficient against the final refined structure as a function of resolution for the initial SAD phases of 90°/270°; (b) after partial resolution of the twofold ambiguities; (c) after adding the S-atom contributions; (d) after five cycles density modification; and (e) after 20 cycles density modification.

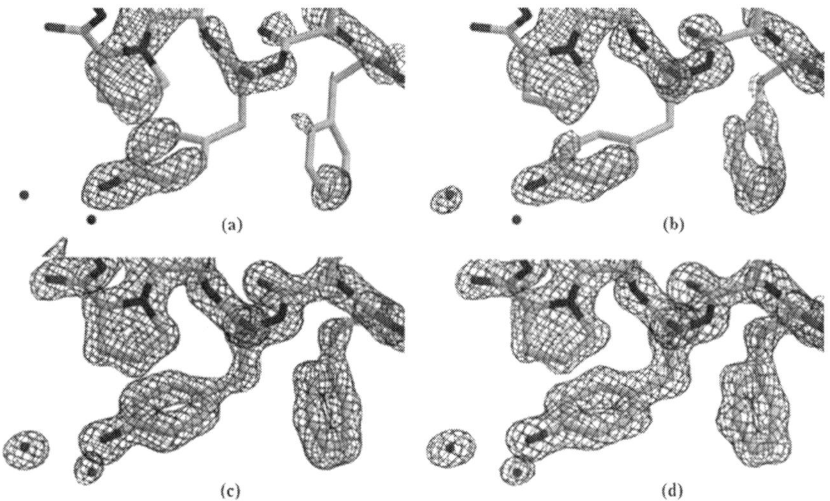

Figure 4. (a–c) are the maps corresponding to parts (a), (b), and (e), respectively of Figure 3. The final *model-free* map (c) is essentially identical to the sigma-A map (d) after the final anisotropic refinement.

8. Examples of phasing and possible improvements

Figure 3 shows the course of phase determination for the cubic insulin test crystal, and Figure 4 shows the corresponding maps. The initial phase information is more reliable at lower resolution where the anomalous differences are more significant, but after 20 cycles of density modification, the procedure has converged with a respectable map CC of 0.97 for the full resolution range. For further in-house sulfur-SAD examples, see Debreczeni et al. [10–12].

When the anomalous signal does not extend to sufficient resolution to resolve disulfides, it has been standard practice to search for *super-sulfur* atoms. An effective alternative [12] is to modify the peak search to locate the best positions for S–S units in the slightly elongated electron density maxima. These *resolved disulfides* not only improve the performance of the substructure solution, but also give a much better phase extension to higher resolution and better final map CC s. The CPU time overhead is negligible.

This suggests that searching for small fragments in the real space part of the dual-space recycling may be a good way of extending direct methods to lower resolution, provided that it can be done efficiently. Patterson seeding with larger heavy atom fragments (e.g., Fe_2S_2, Fe_4S_4, and $Ta_6Br_{12}{}^{2+}$ clusters) has been incorporated into a test version of SHELXD and looks promising.

SHELXE was designed to be fast, robust, and easy to use, as a contribution to *high-throughput* phasing. Despite its lack of sophistication, there are three sets of circumstances under which high quality *model-free* maps are obtained directly:

- When high-quality MAD data are used
- When the solvent content is high (>0.6)
- When the resolution of the native data is very high (<1.5 Å)

These maps have a surprisingly atomic appearance, and show a tendency for the features corresponding to lower electron density (e.g., disordered side chains) to be suppressed. This may be useful for tracing an initial model, but could be a disadvantage for interpreting the finer details of disorder.

9. Advantages of SAD phasing for high-throughput pipelines

Despite the fact that a SAD experiment produces less phase information than MAD or SIRAS, there are several advantages that make it attractive for highly automated pipelines. The wavelength must still be appropriate, for example, 1.54 Å is good for Co ($f'' = 3.6$) but poor for Zn ($f'' = 0.68$), but it is much less critical. Radiation damage is less of a problem and there is no problem with isomorphism; only one crystal is needed. If the crystal diffracts well enough, it may be a good strategy to collect the anomalous data to modest resolution with a high redundancy (e.g., using a kappa goniometer) in-house, then to collect a low redundancy dataset to the highest possible resolution (possibly using the same frozen crystal) on a synchrotron. This minimizes the use of expensive beam time and gets around the problem that many European beam lines have only a single rotation axis.

10. Scripts and GUIs

The programs SHELXC, SHELXD, and SHELXE are called by a number of GUIs and pipelines, including HKL2MAP [13] that was designed specially for use with them and has the advantage that it produces very educational graphical displays of the various numerical information output by the programs. Since these programs are fast, robust, and require a minimum of starting information they are particularly suitable for incorporation into beamline pipelines, e.g., to see if a structure is soluble before the crystal is taken off the beamline. The programs may also be run using a simple UNIX script as for the solution of the structure of elastase by sulfur-SAD illustrated in the following example. SHELXC reads the native data (containing Friedel opposites) and sets up the files needed for substructure solution with SHELXD and SAD phasing and density modification with SHELXE. This script searches for 12 sulfur atoms with a minimum distance apart of 1.7 Å (four disulfide bonds are present). At lower resolution the combination of FIND 8, MIND 3.5, and DSUL 4 might be better to look for the four disulfides and four other sulfurs. 20 density modification cycles are performed for each heavy atom enantiomorph and the solvent content is 0.37.

```
shelxc elas << EOF
SAD elastase.sca
CELL 49.704 57.895 74.169 90 90 90
SPAG P212121
FIND 12
MIND -1.7
EOF
shelxd elas_fa
shelxe elas elas_fa -s0.37 -m20
shelxe elas elas_fa -s0.37 -m20 -i
```

Acknowledgements

I am grateful to Isabel Usón, Thomas R. Schneider, and Tim Grüne for many useful discussions, and to the Fonds der Chemischen Industrie and the EU BIOXHIT project for support.

References

1. Karle, J. (1980) Some developments in anomalous dispersion for the structural investigation of macromolecular systems in biology. *International Journal of Quantum Chemistry Symposium*, **7**: 357–367.
2. Hendrickson, W.A., Smith, J.L., and Sheriff, S. (1985) Direct phase determination based on anomalous scattering. *Methods in Enzymology*, **115**, 41–55.
3. Sheldrick, G.M. (2002) Macromolecular phasing with SHELXE. *Zeitschrift für Kristallographie*, **217**: 644–650.
4. Hendrickson, W.A., and Lattman, E.E. (1970) Representation of phase probability distributions for simplified combination of independent phase information. *Acta Crystallographica*, **B26**: 136–143.
5. Miller, R., Gallo, S.M., Khalak, H.G., and Weeks, C.M. (1994) SnB: crystal structure determination via Shake-and-Bake. *Journal of Applied Crystallography*, **27**: 613–621.
6. Usón, I., and Sheldrick, G.M. (1999) Advances in direct methods for protein crystallography. *Current Opinion in Structural Biology*, **9**: 643–648.
7. Grosse-Kunstleve, R.W. and Adams, P.D. (2003) Substructure search procedures for macromolecular direct methods. *Acta Crystallographica*, **D59**: 1966–1973.
8. Sheldrick, G.M., Hauptman, H.A., Weeks, C.M., Miller, M., and Usón, I. (2001) Direct methods: ab initio phasing. *In International Tables for Crystallography*, vol. F. Edited by Arnold, E., and Rossmann, M. Dordrecht: Kluwer Academic, pp. 333–351.

9. Schneider, T.R., and Sheldrick, G.M. (2002) Substructure solution with SHELXD. *Acta Crystallographica*, **D58**: 1772–1779.
10. Debreczeni, J.É., Bunkóczi, G., Girmann, B., and Sheldrick, G.M. (2003) In-house determination of the lima bean trypsin inhibitor: a low-resolution sulfur-SAD case. *Acta Crystallographica*, **D59**: 393–395.
11. Debreczeni, J.É., Bunkóczi, G., Ma, Q., Blaser, H., and Sheldrick, G.M. (2003) In-house measurement of the sulfur anomalous signal and its use for phasing. *Acta Crystallographica*, **D59**: 688–696.
12. Debreczeni, J.É., Girmann, B., Zeeck, A., Krätzner, R., and Sheldrick, G.M. (2003) Structure of viscotoxin A3: disulfide location from weak SAD data. *Acta Crystallographica*, **D59**: 2125–2132.
13. Pape, T. and Schneider, T.R. (2004) HKL2MAP: a graphical user interface for macromolecular phasing with SHELX programs. *Journal of Applied Crystallography*, **37**: 843–844.
14. Yao, J.-X. (2002) ACORN in CCP4 and its applications. *Acta Crystallographica*, **D58**: 1941–1947.
15. Terwilliger, T.C. (1999) σ^2_R, a reciprocal-space measure of the quality of macromolecular electron-density maps. *Acta Crystallographica*, **D55**: 1174–1178.

LIKELIHOOD-BASED EXPERIMENTAL PHASING IN *PHASER*

AIRLIE J. MCCOY, LAURENT C. STORONI,
AND RANDY J. READ

*University of Cambridge, Department of Haematology,
Cambridge Institute for Medical Research, Wellcome
Trust/MRC Building, Hills Road, CB2 0XY, UK*

Abstract: There are two likelihood functions in *Phaser* for use in experimental phasing: one for MIR/MIRAS/MAD phasing, and one specially developed for SAD (single-wavelength anomalous dispersion) phasing. The MIR/MIRAS/MAD function involves a two-dimensional (2D) integration over the complex plane. 2D integration is a slow process, and in the course of a typical experimental phasing run needs to be performed millions of times. We review here how both likelihood functions are derived, and discuss methods for overcoming the computational bottleneck in the integration of the MIR/MIRAS/MAD function, as implemented in the program *Phaser*.

Keywords: experimental phasing; maximum likelihood; Phaser.

1. Introduction

Experimental phasing by MIR (multiple-wavelength isomorphous replacement), MIRAS (multiple-wavelength isomorphous replacement with anomalous scattering), and MAD (multiple-wavelength anomalous dispersion) currently use the same likelihood function [1–4]. This likelihood function has the least approximations when used for MIR. The approximation is much worse when used for MIRAS and MAD, because correlations between F^+ and F^- and the errors are not properly accounted for. The function is far from ideal, but experience shows that it is still capable of successfully phasing many structures.

Recently, a second likelihood function has been developed for phasing by SAD (single-wavelength anomalous dispersion). This likelihood function explicitly accounts for the correlations between F^+ and F^- [5, 6]. Because the correlations are accounted for, better phases can be obtained using this function than by taking SAD as just a "special" case of the MIR/MIRAS/MAD likelihood function.

67

R.J. Read and J.L. Sussman (eds.): Evolving Methods for Macromolecular Crystallography, 67–77
© 2007. *Springer*

This paper describes the theory behind the MIR/MIRAS/MAD and SAD likelihood functions, and the way they are implemented in the program *Phaser*. The basic theory of the likelihood functions is similarly covered, but in more detail and illustrated using games of dice, in McCoy [7].

2. Basic concepts

Likelihood functions for experimental phasing are derived using the four basic concepts of maximum likelihood, independence, log-likelihood, and the central limit theorem.

2.1. MAXIMUM LIKELIHOOD

The likelihood is the probability of the data given the model. In crystallography, the data are the observed structure factor amplitudes F_O, so in order to compare like with like, the model must be represented by the calculated structure factor amplitudes F_C.

$$P(\text{data};\text{model}) = P(F_O; F_C)$$

It will be shown below that the likelihood function for F_O given F_C for a single (acentric) reflection and a single derivative for MIR phasing is given by a function called a *Rice distribution* [8, 9].

$$P(F_O; F_C) = \frac{2F_O}{\sigma_\Delta^2 + \sigma_F^2} e^{-\left[\frac{F_O^2 + D^2 F_C^2}{\sigma_\Delta^2 + \sigma_F^2}\right]} I_0 \left|\frac{2F_O D F_C}{\sigma_\Delta^2 + \sigma_F^2}\right| \tag{1}$$

$$\equiv \Re(F_O, D F_C, \sigma_\Delta^2 + \sigma_F^2)$$

where I_0 is the modified Bessel function of order zero. The terms D and σ_Δ describe the errors in the model and σ_F describes the errors in the data. From now onwards, Rice functions will be referred to by the function name $\Re$. Similarly, one-dimensional (1D) and 2D-Gaussian functions (bell-shaped curves) will be referred to the function name $\mathcal{G}$ (bold font for 2D-Gaussian).

$$\mathcal{G}\left(F_O, F_C, 2\sigma^2\right) = \frac{1}{\sqrt{2\pi\sigma^2}} e^{-\left(\frac{(F_O - F_C)^2}{2\sigma^2}\right)} \text{ and } \boldsymbol{\mathcal{G}}\left(\mathbf{F}_O, \mathbf{F}_C, \sigma^2\right) = \frac{1}{\pi\sigma^2} e^{-\left(\frac{|\mathbf{F}_O - \mathbf{F}_C|^2}{\sigma^2}\right)}$$

where $\mathbf{F}_O$ and $\mathbf{F}_C$ are the observed and calculated structure factor vectors. Here $F_C/\mathbf{F}_C$ is the mean of the Gaussian ($\mathcal{G}/\boldsymbol{\mathcal{G}}$, respectively) and the variance is σ^2.

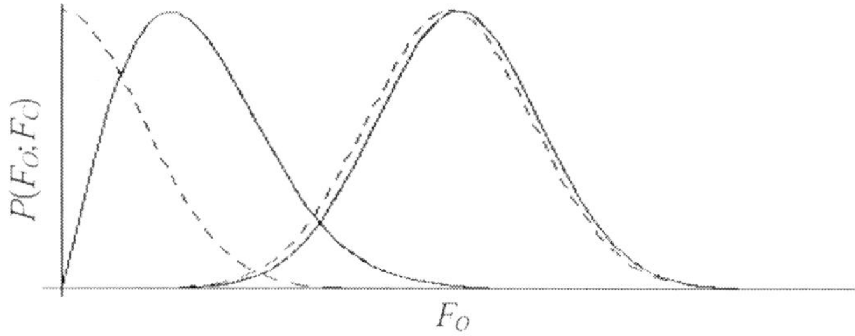

Figure 1. Rice functions (*solid lines*) and Gaussian functions (*dashed lines*) shown for the same two values of F_C. The plots with maxima closest to the origin have F_C of zero, and the plots with maxima distant from the origin have non-zero F_C.

The Rice function looks like a 1D-Gaussian when F_C is large compared to the errors. (This is why old crystallographic probability functions, which used least squares functions, were quite accurate for large values of F_C). However, when F_C is small (and the maximum of the Rice function is close to the origin), the shape of the plot becomes distorted from a Gaussian – it looks like a Gaussian that has been "squashed" against the origin (Figure 1).

The Rice distribution is the most important distribution for maximum likelihood in crystallography, and it is ubiquitous in maximum likelihood functions used for molecular replacement, refinement and experimental phasing (see McCoy [7] for a review).

2.2. INDEPENDENCE AND LOG-LIKELIHOOD

Reflections are assumed to be independent, even though to a certain extent they are not. Correlations between reflections are introduced by the presence of solvent and by any non-crystallographic symmetry. In general, correlations make the determination of the likelihood difficult, if not impossible, and it is much simpler to ignore them. In the case of crystallography the correlations between reflections are sufficiently weak that the approximation of assuming independence is very good. (However, these correlations are the basis of phasing by direct methods, and so are vital to this alternative phasing technique). With the assumption of independence, the total likelihood becomes the product of reflection likelihoods. For reasons of numerical stability, in practice, the log-likelihood is calculated rather than the likelihood. To calculate the total log-likelihood for all the reflections, the sum of the log-likelihoods for each reflection is used. Since optimization routines are designed to minimize functions, the best model parameters (e.g., heavy atom positions, occupancies, and B-factors) are found by minimizing the −log-likelihood.

2.3. INTEGRATING OUT NUISANCE VARIABLES

It may be easier to develop a probability function using an extra variable than to attempt to develop it without ever referring to the extra variable. These extra variables are called "nuisance" variables, and at the end of the analysis are eliminated by integration. Although termed "nuisance", extra parameters can be very useful in probability distributions. In crystallography, it is easier to model the errors in a probability distribution of F_C and F_O in terms of the phased structure factors $\mathbf{F}_C$ and $\mathbf{F}_O$, rather than in terms of the structure factor amplitudes alone. The introduced variable, the phase difference α, is a nuisance variable (a case where a nuisance variable is very useful) and must be integrated out of the probability distribution at the end of the analysis.

$$P(F_O;F_C)= \int\limits_{0}^{2\pi} P(F_O,\alpha;F_C)\,d\alpha \qquad (2)$$

2.4. CENTRAL LIMIT THEOREM

The central limit theorem (historically called the "law of errors") states that the distribution of the average tends to be Gaussian, even when the distributions of the individual terms from which the average is computed are decidedly non-Gaussian. In crystallography, the central limit theorem allows us to describe the errors in the structure factors (in reciprocal space) that arise from errors in the atomic model (in real space). It says that, even though the errors in an individual atom's contribution to the total structure factor may be very complicated, in the end, the error for the total structure factor (the sum of the atomic structure factor contributions) is a simple 2D-Gaussian in reciprocal space. (Note that the sum and the average of a set of values is the same except for dividing by the number of values, so the central limit theorem applies equally to the average and to the sum of a set of variables.)

Consider an atom in the unit cell. If the atom at its average position with its average scattering has a structure factor f, then variation in the atom's position corresponds to variation in the phase of f and variation in the scattering corresponds to variation in the length of f. The resulting distribution of the structure factor (due to variation in the atom's position and scattering) is boomerang shaped (Figure 2), running around the circumference of the circle of radius $|f|$. The distribution is symmetric about f so the average structure factor is shorter than f by a fraction d (d between 0 and 1), and in the same direction as f (i.e., df). One way to see that the average of the distribution is df is to imagine where you would have to put your finger to balance a boomerang on your fingertip so that it was lying horizontally. You would have to put your finger not at

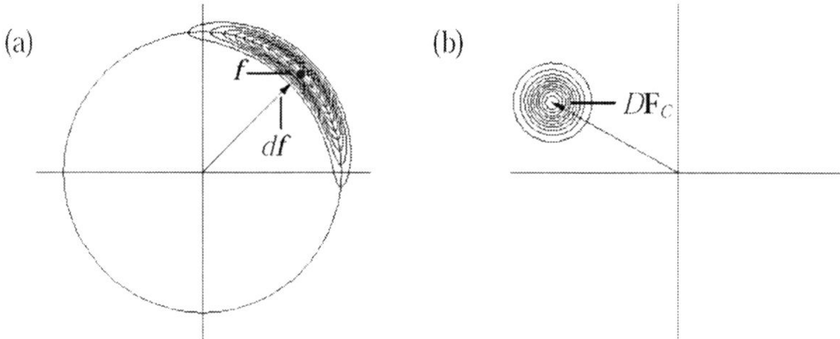

Figure 2. (a) Boomerang-shaped contours for the probability distribution for the structure factor for a single atom, accounting for errors in position and scattering of the atom. The average structure factor is in the same direction as the structure factor for the atom with its average position and scattering *f*, but shortened by a factor *d* (*d* between 0 and 1). (b) After adding the atomic structure factors and their errors, the final distribution is a 2D-Gaussian centred on $D\mathbf{F}_C$ (*D* between 0 and 1).

the centre of the boomerang, but somewhere closer to the edge of the inner curve of the boomerang – or off the edge of the boomerang altogether, if that were possible.

Now consider all the atoms in the unit cell. The errors in all the individual structure factors are boomerang-shaped distributions about all the individual *f* vectors. When these atomic structure factor contributions and their errors are summed to give the total structure factor and its error for a given reflection, by the central limit theorem, the resulting distribution is a 2D-Gaussian in reciprocal space. This Gaussian is centred on $D\mathbf{F}_C$ (*D* between 0 and 1), because all the contributing *f* vectors are systematically shortened by fractions *d* (*d* between 0 and 1). The variance of the Gaussian is termed σ_Δ^2.

$$P(\mathbf{F}_O; \mathbf{F}_C) = \mathcal{G}\left(\mathbf{F}_O, D\mathbf{F}_C, \sigma_\Delta^2\right)$$

The relationship between $P(F_O, \alpha; F_C)$ and $P(\mathbf{F}_O; \mathbf{F}_C)$ is given by

$$P(F_O, \alpha; F_C) = F_O \times P(\mathbf{F}_O; \mathbf{F}_C) \tag{3}$$

where the factor F_O is introduced by changing the description of the Fs from Cartesian coordinates (i.e., expressed in terms of real and imaginary components) to polar coordinates (i.e., expressed in terms of radial and angular components; this factor is called the *Jacobian*). Using equation 3 in equation 2 gives a Rice function (for more details see McCoy [7])

$$P(F_O; F_C) = \Re\left(F_O, DF_C, \sigma_\Delta^2\right)$$

There are also experimental errors (σ_F), which are accounted for by widening the probability distribution, a method that is termed *inflating the variance* [4, 10, 11]. The Rice distribution including experimental errors is given by equation 1.

3. MIR/MIRAS/MAD likelihood

When considering how to phase a new crystal structure by experimental methods, a choice is usually made of whether to attempt structure solution by MIR/MIRAS or MAD. The distinction is drawn in part because MAD requires a tunable wavelength source (synchrotron), and seleno-methionine protein is often produced expressly for the purpose. However, despite the differences in experimental technique between the two, the underlying theory is based on the theory for phasing by MIR, and then approximations used to expand the function so that it can account for phasing by MIRAS and MAD.

3.1. MIR LIKELIHOOD

The likelihood for a reflection r in MIR is the probability of the set of F_O (denoted $\{F_{Oj}\}$), given the set of $\mathbf{F}_{H}$ (denoted $\{\mathbf{F}_{Hj}\}$), for the derivatives j. The "native" data set (if any) is not treated as "special". It is just treated as another derivative, with no heavy atoms.

$$P\text{-}\mathrm{MIR_R} = P\left(\{F_{Oj}\}; \{\mathbf{F}_{Hj}\}\right)$$

There are significant correlations in the data because all data sets share the scattering from the native protein component, i.e., if a reflection is strong or weak in the native, then it is likely to be strong or weak in all the derivative data sets as well. To simplify the analysis a (useful) nuisance variable is introduced, the "true" structure factor $\mathbf{F}_T$, which is the component of scattering shared by the native and derivatives [12]. The introduced (useful) nuisance variable $\mathbf{F}_T$ must be integrated out of the probability distribution at the end of the analysis (a double integral over the complex plane)

$$P\left(\{F_{Oj}\}; \{\mathbf{F}_{Hj}\}\right) = \int_{-\infty}^{+\infty} \int_{-\infty}^{+\infty} P\left(\{F_{Oj}\}, \mathbf{F}_T; \{\mathbf{F}_{Hj}\}\right) d\mathbf{F}_T \tag{4}$$

The reason for introducing the nuisance variable $\mathbf{F}_T$ is that, by explicitly including the correlated component of the scattering between all the data, the "left over" parts of the scattering can be taken as independent. Therefore, the probabilities $P(F_{Oj}; \mathbf{F}_T, \mathbf{F}_{Hj})$ are (approximately) independent and can be multiplied to give $P(\{F_{Oj}\}, \mathbf{F}_T; \{\mathbf{F}_{Hj}\})$. With some approximations and manipulations of the probabilities (see McCoy [7] for more explanation),

$$P\left(\left\{F_{Oj}\right\}, \mathbf{F}_T; \left\{\mathbf{F}_{Hj}\right\}\right) \approx \prod_{j=1}^{J} P(F_{Oj}; \mathbf{F}_{Hj}, \mathbf{F}_T)$$

If the heavy atom model is perfect (and thus $\mathbf{F}_H$ is perfect), and the protein component of the derivative is identical (isomorphous) to $\mathbf{F}_T$, then $\mathbf{F}_C = \mathbf{F}_T + \mathbf{F}_H$. However, $\mathbf{F}_H$ will not be perfect because the heavy atoms will not have perfect positions and occupancies and some of the sites may be missing from the model, and $\mathbf{F}_T$ will not be perfectly isomorphous with the protein component of the derivative. Errors in the position and scattering of the heavy atoms lead to boomerang-shaped errors in the individual atomic structure factors, but, by the central limit theorem, the total error in $\mathbf{F}_{Hj}$ is a 2D-Gaussian centred on $D_{Hj}\mathbf{F}_{Hj}$ (D_{Hj} between 0 and 1). Refining the D_{Hj} has the same effect as refining the occupancies and B-factors of the heavy atoms, and so can be absorbed by these parameters during refinement. Similarly, non-isomorphism errors mean that the total error in $\mathbf{F}_T$ is a 2D-Gaussian centred on $D_j\mathbf{F}_T$ (D_j between 0 and 1), but this D cannot be absorbed by other parameters. Including these errors, the calculated structure factor is given by

$$\mathbf{F}_{Cj} = D_j\mathbf{F}_T + \mathbf{F}_{Hj}, \quad \text{so that } F_{Cj} = |D_J\mathbf{F}_T + \mathbf{F}_{Hj}|$$

The trick to deriving a maximum likelihood MIR function is to introduce yet another nuisance parameter, the phase difference α between $\mathbf{F}_{Oj}$ and $\mathbf{F}_{Cj}$. The probability of $\mathbf{F}_{Oj}$ is a 2D-Gaussian in reciprocal space centred on $\mathbf{F}_{Cj}$ with variance $\sigma_{\Delta j}^2$. Integrating out the phase between $\mathbf{F}_{Oj}$ and $\mathbf{F}_{Cj}$ and including experimental error by inflating the variance gives a Rice distribution.

$$P(F_{Oj}; \mathbf{F}_{Hj}, \mathbf{F}_T) = \Re\left(F_{Oj}, |D_j\mathbf{F}_T + \mathbf{F}_{Hj}|, \sigma_{\Delta j}^2 + \sigma_{Fj}^2\right) \tag{5}$$

Substituting equation 5 into equation 4, and converting $\mathbf{F}_T$ to polar coordinates (introducing the Jacobian, in this case F_T),

$$P\text{-MIR}_r = \int_0^{2\pi} \int_0^{\infty} \prod_{j=1}^{J} \Re(F_{Oj}, |D_j\mathbf{F}_T + \mathbf{F}_{Hj}|, \sigma_{\Delta J}^2 + \sigma_{Fj}^2) F_T \, dF_T \, d\alpha_T \tag{6}$$

Unfortunately, integrating out $\mathbf{F}_T$ from equation 6 cannot be done analytically: it must be done numerically. This can be done by calculating values of the function over a 2D grid of phases α_T and amplitudes F_T, and summing the values. If the phases are sampled at intervals of 6 degrees (for example), and amplitudes at 5 points around F_O (for example), integrating the function this way requires 300 Rice function evaluations per derivative per integral. This is a computationally intensive process, and as it needs to be performed millions of times in the course of a typical refinement, it is important to make the integration algorithm very fast.

One way to speed up the 2D integration is to be selective in the points that are calculated for inclusion in the sum – only those points where the function values are expected to be significant are included. The sampling interval for the phase α_T can be altered depending on the variances of the Rice functions contributing to the product in equation 6. If the variances are large, the function will not change rapidly and the phase sampling can be coarser. If the variances are small and the Rice functions sharper, then more phase points need to be included in the summation. Similar selectivity can be applied to points for inclusion in the amplitude part of the integration.

Another way to speed up the integration would be to approximate the function with one that *can* be integrated analytically. Any function used to approximate the Rice function should differ from the true Rice function less than the other errors present, and to be advantageous, the integral of the approximation should be a numerically simple function to calculate.

3.2. MIRAS LIKELIHOOD

The likelihood for a reflection r for MIRAS is the probability of the set of F_O^+ and F_O^- for all derivatives j (denoted $\{F_{Oj}^+, F_{Oj}^-\}$) given the set of $\mathbf{F}_H^+$ and $\mathbf{F}_H^-$ for all derivatives j (denoted $\{\mathbf{F}_{Hj}^+, \mathbf{F}_{Hj}^-\}$), rather than just the mean F_{Oj} and mean $\mathbf{F}_{Hj}$ as for MIR.

$$P\text{-MIRAS}_r = P\left(\left\{F_{Oj}^+, F_{Oj}^-\right\}; \ \left\{\mathbf{F}_{Hj}^+, \mathbf{F}_{Hj}^-\right\}\right)$$

This probability function is difficult to generate because F_O^+ and F_O^- are highly correlated (if F_O^+ is large or small, F_O^- will also be large or small). This problem is partially avoided if the mean F_O and anomalous difference ΔF_O are used instead of F_O^+ and F_O^-, as these are less correlated with one another (if the mean F_O is large, the anomalous difference ΔF_O need not be large) [4, 13, 14]. The probabilities for the mean F_O and ΔF_O are then considered independent. The probability for the mean F_O is the probability derived above (equation 6) for the MIR function. The probability for ΔF_O is approximated by a 1D-Gaussian function. Following the same method as used for deriving the expression for $P\text{-MIR}_r$,

$$P\text{-MIRAS}_r = \int_0^{2\pi} \int_0^{\infty} \prod_{J=1}^{J} \mathcal{G}\left(\Delta F_{Oj}, \Delta F_{Cj}, 2\sigma_{\alpha j}^2 + \sigma_{\Delta Fj}^2\right)$$
$$\Re\left(F_{Oj}, F_{Cj}, \sigma_{\Delta j}^2 + \sigma_{Fj}^2\right) F_T \, dF_T \, d\alpha_T$$

$$\text{where } \Delta F_{Cj} = \left| \, \left\| D_j \mathbf{F}_T + \mathbf{F}_{Hj}^+ \right\| - \left\| D_j \mathbf{F}_T + \mathbf{F}_{Hj}^- \right\| \, \right| \tag{7}$$

3.3. MAD LIKELIHOOD

The likelihood function for MAD is the probability of the set of $F_{O\lambda}^+$ and $F_{O\lambda}^-$ given the set of $\mathbf{F}_{H\lambda}^+$ and $\mathbf{F}_{H\lambda}^-$ for wavelengths λ. This is equivalent to P-MIRAS$_r$ with the wavelengths λ corresponding to derivatives j, and current MAD phasing methods simply take MAD as a special case of MIRAS. However, the derivation of both the Gaussian and Rice approximations of P-MIRAS$_r$ assumed that errors in the models of heavy atoms between derivatives (i.e., wavelengths) were uncorrelated with one another. This is definitely not the case in MAD; since the heavy atom model is the same in all "derivatives", the errors *must* be correlated. To remove these correlations, nuisance parameters corresponding to each unknown phase must be introduced, in the same way that the "true" $\mathbf{F}$ was introduced to account for the correlations in the MIR case. Removal of the nuisance phases then requires one integration per phase at the end of the analysis. For example, in 2λ-MAD, a four-dimensional (4D) integration over α_1^+, α_1^-, α_2^+, and α_2^- is required. One of these integrations can always be done analytically (to give a Rice distribution), but the others must be done numerically. As the dimensionality of the integration increases the computational time increases with the power of the dimensionality, but a more intractable problem is that the calculation becomes numerical unstable, i.e., small differences in input get magnified as the calculation proceeds, and the algorithm produces in effect "random" numbers. Both speed and stability issues may be addressed in the future by finding approximations to the integrals that can be calculated analytically.

4. SAD likelihood

The likelihood for SAD is the probability of F_O^+ and F_O^- given $\mathbf{F}_H^+$ and $\mathbf{F}_H^-$.

$$P\text{-SAD}_r = P\left(F_O^+, F_O^-; \mathbf{F}_H^+, \mathbf{F}_H^- \right)$$

F_O^+ and F_O^- are highly correlated, as already discussed. To remove the correlations it is useful to introduce phases α^+ and α^- of F_O^+ and F_O^-, which must be integrated out at the end of the analysis.

$$P\text{-SAD}_r = \int_0^{2\pi} \int_0^{2\pi} P\left(F_O^+, \alpha^+, F_O^-, \alpha^-; \mathbf{F}_H^+, \mathbf{F}_H^- \right) d\alpha^+ \, d\alpha^-$$

The probability for $P\text{-SAD}_r$ can be rearranged using the basic probability identity $P(B, A; C) = P(A; C) \times P(B; C, A)$

$$P\left(F_O^+, \alpha^+, F_O^-, \alpha^-; \mathbf{F}_H^+, \mathbf{F}_H^-\right) = P\left(F_O^-, \alpha^-; \mathbf{F}_H^+, \mathbf{F}_H^-\right) P\left(F_O^+, \alpha^+; F_O^-, \alpha^-, \mathbf{F}_H^+, \mathbf{F}_H^-\right)$$

The advantage of this formula is that only the second probability involves α^+ (although they both involve α^-). The integration over α^+ is thus restricted to this part of the probability function

$$P\text{-SAD}_r = \int_0^{2\pi} P(F_O^-, \alpha^-; \mathbf{F}_H^+, \mathbf{F}_H^-) \left\{ \int_0^{2\pi} P\left(F_O^+, \alpha^+; F_O^-, \alpha^-, \mathbf{F}_H^+, \mathbf{F}_H^-\right) d\alpha^+ \right\} d\alpha^-$$

The integration in curly brackets gives a Rice distribution [5]. The remaining probability function is a 2D-Gaussian centred on $\mathbf{F}_H^-$. The Rice distribution accounts for the anomalous difference and the 2D-Gaussian term accounts for contribution of the heavy atoms to the normal scattering. Only a single numerical (phase) integration is required.

$$P\text{-SAD}_r = F_O^- \int_0^{2\pi} \mathcal{G}\left(\mathbf{F}_O^-, \mathbf{F}_H^-, \sigma_\Delta^2 + \sigma_{F^-}^2\right) \Re\left(F_O^+, F_C^+, \sigma_+ + \sigma_{F^+}^2 + \sigma_{F^-}^2\right) d\alpha^-$$

where $F_C^+ = \left| \mathbf{F}_H^+ + \mathbf{D}_\phi(\mathbf{F}_O^- - \mathbf{F}_H^-) \right|$

5. Discussion

The approximations made in developing the experimental phasing likelihood functions leave much room for improvement. In particular, taking account of the correlations between structure factors and errors will lead to better phases in marginal cases. The correct way to handle these correlations is to introduce a nuisance parameter for each correlation, and then integrate out these parameters from the equation at the end of the analysis. These more sophisticated analyses depend on utilizing future increases in computing power. However, even with increases in computer power it will still be necessary to find numerically stable formulations of the integrations.

Experimental phasing methods described here have been implemented in the program *Phaser*. *Phaser* is available through the *Phenix* [15] and *CCP4* [16] software distributions, and directly from the authors (for further details see http://www-structmed.cimr.cam.ac.uk/phaser).

References

1. Bricogne, G. (1991) A maximum-likelihood theory of heavy-atom parameter refinement in the isomorphous replacement method. In *Isomorphous Replacement and Anomalous Scattering: Proceedings of the CCP4 Study Weekend.* Edited by Evans, P.R. and Leslie, A.G.W. Warrington: Daresbury Laboratory, pp. 60–68.
2. Read, R.J. (1991) In *Isomorphous Replacement and Anomalous Scattering: Proceedings of the CCP4 Study Weekend.* Edited by Evans, P.R. and Leslie, A.G.W. Warrington: Daresbury Laboratory, pp. 69–79.
3. Read, R.J. (1994) Lecture Notes from the Workshop on Isomorphous Replacement Methods in Macromolecular Crystallography. *American Crystallographic Association Annual Meeting,* Atlanta, GA.
4. La Fortelle, E. de and Bricogne, G. (1997) Maximum-likelihood heavy-atom parameter refinement for multiple isomorphous replacement and multi-wavelength anomalous diffraction methods. *Methods in Enzymology,* **276**: 472–494.
5. McCoy, A.J., Storoni, L.C, and Read, R.J. (2004) Simple algorithm for a maximum-likelihood SAD function. *Acta Crystallographica,* **D60**: 1220–1954.
6. Pannu, N.S., and Read, R.J. (2004) The application of multivariate statistical techniques improves single-wavelength anomalous diffraction phasing. *Acta Crystallographica,* **D60**: 22–27.
7. McCoy, A.J. (2004) Liking Likelihood. *Acta Crystallographica,* **D60**: 2169–2183.
8. Sim, G.A. (1959) The distribution of phase angles for structures containing heavy atoms. II. A modification of the normal heavy-atom method for non-centrosymmetrical structures. *Acta Crystallographica,* **12**: 813–815.
9. Read, R.J. (1990) Structure-factor probabilities for related structures. *Acta Crystallographica,* **D46**: 900–912.
10. Green, E.A. (1979) A new statistical model for describing errors in isomorphous replacement data: the case of one derivative. *Acta Crystallographica,* **A35**: 351–359.
11. Murshudov, G.N., Vagin, A.A., and Dodson, E.J. (1997) Refinement of macromolecular structures by the maximum-likelihood method. *Acta Crystallographica,* **D53**: 240–255.
12. Read, R.J. (2003) New ways of looking at experimental phasing. *Acta Crystallographica,* **D59**: 1891–1902.
13. North, A.C.T. (1965) The combination of isomorphous replacement and anomalous scattering data in phase determination of non-centrosymmetric reflexions. *Acta Crystallographica,* **18**: 212–216.
14. Matthews, B.W. (1966) The extension of the isomorphous replacement method to include anomalous scattering measurements. *Acta Crystallographica,* **20**: 82–86.
15. Adams, P.D., Grosse-Kunstleve, R.W., Hung, L.-W., Ioerger, T.R., McCoy, A.J., Moriarty, N.W., Read, R.J., Sacchettini, J.C., Sauter, N.K., and Terwilliger, T.C. (2002) PHENIX: building new software for automated crystallographic structure determination. *Acta Crystallographica,* **D58**: 1948–1954.
16. Collaborative Computing Project 4 (1994) The *CCP*4 suite: programs for protein crystallography *Acta Crystallographica,* **D50**: 760–763.

STOCHASTIC MOLECULAR REPLACEMENT

NICHOLAS M. GLYKOS

*Department of Molecular Biology and Genetics, Democritus
University of Thrace, Dimitras 19, 68100 Alexandroupolis,
Greece*

Abstract: Classical molecular replacement methods and the newer six-dimensional searches treat molecular replacement as a succession of subproblems of reduced dimensionality. Due to their divide-and-conquer approach, these methods necessarily ignore (at least during their early stages) the very knowledge that a target crystal structure may comprise, for example, more than one copy of a search model, or, several models of different types. Here, we describe a stochastic multidimensional molecular replacement algorithm that represents one of the most general formulations and methods of solution of the molecular replacement problem. The method uses all information available at hand for a given problem (both with respect to the number and type of the search models, and the measured crystallographic data) without resorting to Patterson function-based heuristics. The algorithm only demands that for the given search models the true crystal structure is the one for which the agreement between the observed and calculated structure factor amplitudes is maximised. This approach has been shown to be capable of successfully locating solutions even in cases as complex as a 23-dimensional, four-body search.

Keywords: stochastic molecular replacement; multidimensional molecular replacement; reverse Monte Carlo; simulated annealing.

1. Introduction

One of the most direct approaches to the solution of the molecular replacement problem is based on the following premise: determine the orientations and positions (of the search models) for which the agreement between the observed and calculated structure factor amplitudes is maximised. If the structures of the search models are sufficiently accurate, these orientations and positions will correspond to the correct molecular replacement solution.

R.J. Read and J.L. Sussman (eds.): Evolving Methods for Macromolecular Crystallography, 79–90
© 2007. *Springer*

More formally, if there are n search models in the asymmetric unit of the target crystal structure, in general, there are $6n$ parameters whose values are to be determined by molecular replacement (three rotational and three translational parameters for each of the search models). These $6n$ parameters, in turn, define a $6n$-dimensional configurational space in which each and every point corresponds to a possible configuration for the target crystal structure, and so, for each and every points it is possible to calculate the value of a suitable statistic (like the R-factor, or the linear correlation coefficient) measuring the agreement between the experimentally observed and the calculated structure factor amplitudes. By assuming that the correct solution corresponds to the global optimum of this statistic, the molecular replacement problem is reduced to one of the unconstrained global optimisation of the chosen statistic in the $6n$-dimensional space defined by the rotational and translational parameters of the molecules. Stated in simpler terms, the aim of this approach is to find which combination of positions and orientations of the n molecules optimises the value of the R-factor or correlation coefficient between the observed and calculated data. In this respect (and by performing the search in a continuous parameter space), the method views molecular replacement as a generalised rigid-body refinement problem.

The most obvious (and possibly least efficient) method for solving the molecular replacement problem using the aforementioned formulation is a systematic (exhaustive) search: for each and every unique combination of orientations and translations of the search models calculate the value of the target function (e.g., the linear correlation coefficient). The combination that optimises the value of the target function is the sought solution. The ever-increasing computing power available to crystallographers has allowed the once prohibitively expensive systematic six-dimensional searches to be successfully applied to new problems [1–3]. Integration of a multistart local optimisation algorithm with a low resolution coarse-grained six-dimensional search has also been reported recently [4] and was shown to be an effective search method with much lessened CPU time requirements (compared with an exhaustive search). Related molecular replacement methods which include (or are based on) six-dimensional searches have also been reported [5, 6].

Although (and as discussed in the previous paragraph) systematic six-dimensional searches can and have been applied to the solution of the molecular replacement problem, their computational cost is so high that other, more efficient search methods have been proposed. These methods use established global optimisation techniques such as genetic algorithms [7], evolutionary programming [8, 9] or simulated annealing [10–12] to search efficiently the multidimensional parameter space. The principal idea of all these methods is the same: the deterministic point-by-point search is given up and the global optimum of the target function is sought by establishing an efficient non-deterministic search path in the parameter space (see next paragraph for a detailed presentation of one of these methods). The most well known and widely used representative of this family of methods is the one based on an evolutionary search algorithm [8, 9] and encoded in the freely available program EPMR.

Of these stochastic molecular replacement methods the most ambitious (and, in terms of computational requirements, expensive) algorithm is the simulated annealing approach that we described in a series of papers [10–12]. Not only the method can deal with high dimensionality molecular replacement problems (the corresponding program can with no modification attempt to solve 36-dimensional problems), but it can immediately and directly use all the search models that are known to be present in the asymmetric unit of the target crystal structure. The following sections contain a detailed presentation of the algorithms and applications of this multidimensional, multimodel approach to molecular replacement.

2. The algorithm

The method is based on a modification of the reverse Monte Carlo technique [13, 14] where instead of minimising the quantity χ^2, one minimises any of the following (user-defined) target functions: (i) the conventional crystallographic R-factor; (ii) the quantity $1.0\text{-}Corr(F_o,F_c)$; and (iii) the quantity $1.0\text{-}Corr(F_o^2,F_c^2)$, where $Corr()$ is the linear correlation coefficient function, F_o and F_c are the observed and calculated structure factor amplitudes. To avoid unnecessary repetition and to simplify the discussion that follows, we will hereafter refer only to the R-factor statistic, on the understanding that any of the correlation-based targets can be substituted for it.

The minimisation procedure follows closely the original Metropolis algorithm [15] and its basic steps are outlined below. Random initial orientations and positions are assigned to all molecules present in the crystallographic asymmetric unit of the target structure, and the R-factor $(= R_{old})$ between the observed and calculated structure factor amplitudes is noted. In the first step of the basic iteration, a molecule is chosen randomly and its orientational and translational parameters are randomly altered. The R-factor $(= R_{new})$ corresponding to this new arrangement is calculated and compared with R_{old}: if $R_{new} \leq R_{old}$, then the new configuration is accepted and the procedure is iterated with a new (randomly chosen) molecule. If $R_{new} > R_{old}$ (i.e., if the new configuration results to a worse R-factor), the new configuration is accepted with probability $\exp((R_{old} - R_{new})/T)$ where T is a control parameter which plays a role analogous to that of temperature in statistical mechanical simulations. This probabilistic treatment again relies on the random number generator: if $\exp((R_{old}-R_{new})/T) > \xi$, where ξ is a random number between 0 and 1, the new configuration is accepted and the procedure iterated. If $\exp((R_{old}-R_{new})/T) \leq \xi$ we return to the previous configuration (the one that resulted to a R-factor equal to R_{old}) and reiterate. Given enough time, this algorithm is guaranteed to find the global optimum of the target function [16]. It is worth noting here that strictly speaking, simulated annealing is guaranteed to find the global optimum of the target function only in the case of the so called Boltzmann annealing, for which the temperature $T(k)$ at each step k of the simulation is given by $T(k) = T_o/\log(k)$, where T_o is the starting temperature [16]. Only with this annealing schedule, and with T_o "sufficiently high", is the algorithm guaranteed to find the global optimum of the target function. In this respect, the linear slow-cooling protocols usually employed in crystallographic calculations are more accurately described by the term "simulated quenching" than the conventionally used term "simulated annealing".

By trading computer memory for speed of execution, the CPU time required per iteration of the Monte Carlo algorithm can be made to be only linearly dependent on the number of reflections of the target structure expanded to space group $P1$. This is achieved by calculating (and storing in memory) the molecular transform of the search models before the actual minimisation is started (see Section 2.1 of reference [7] for more details). Additionally, – and in order to avoid a dependence on the number of molecules present in the asymmetric unit of the target structure – the contribution of each molecule to every reflection is also stored in memory, and so, at each iteration we only have to recalculate the contribution from the molecule that is being tested.

Three silent features of the algorithm presented above are worth discussing in more detail. The first is that all configurations are treated as *a priori* equally probable, without reference to whether their packing arrangement is physically and chemically sensible. Although it is in principle possible to include a van der Waals repulsion term in the method (to take into account bad contacts between symmetry-related molecules), this would destroy the ergodicity property of simulated annealing, that is, it will no longer be possible to guarantee that each and every state of the system can be reached within a finite number of moves. This is due to the fact that once an arrangement is found that allows the efficient packing of the search models and their symmetry equivalent in the target unit cell, no further major rearrangements of the molecular configuration will be possible (especially in tightly packed crystal forms) and the minimisation would come to a halt.

A second limitation of the method is that by optimising a global statistic like the correlation coefficient or the R-factor, it tries to simultaneously match both the self vectors (of the search models) and all of the cross vectors (between search models and their crystallographically equivalent molecules). The problem with this approach is that as the search model is becoming worse and worse, the agreement for the cross vectors (which are on the average longer) deteriorates much faster than for the (shorter) self vectors, thus reducing the effective signal-to-noise ratio for the correct solution. In contrast, traditional rotation function (possibly due to restricting itself to a self-vector-enriched volume of the Patterson function) is expected to be able to sustain a recognisable solution even for quite inaccurate starting models, increasing in this way the probability that a subsequent translation function will also be successful. The implication of this analysis is that when a sufficiently accurate search model is not available, then this stochastic method may be less sensitive (compared with the conventional Patterson-based methods) in identifying the correct solution.

The third (and most important) limitation of this method is that by treating the problem as $6n$-dimensional, it ignores all the information offered by the properties of the Patterson function. This includes information about the probable orientations of the molecules (usually presented in the form of the cross-rotation function) and of the relationships between them (usually in the form of the self-rotation function). The method as described above also fails to automatically take into account cases of purely translational non-crystallographic symmetry [17], although it is relatively easy to account for such forms of non-crystallographic symmetry through the incorporation of additional fixed symmetry elements. It is worth mentioning here that if the assumption of topological segregation of the self and cross vectors in the Patterson function holds, then molecular replacement problems are not $6n$-dimensional, but rather two $3n$-dimensional problems: the first is a generalised cross-rotation function which would attempt to determine the orientation of all n molecules simultaneously (by taking into account not only the agreement between the observed Patterson function and an isolated set of self vectors from just one of the search models, but also the interactions between the n copies of self-vector sets that are necessarily present in the observed Patterson function). The second is a generalised translation function which would attempt to simultaneously determine the positions of all n properly oriented (from the first step) search models. For this reason – and as long as the assumptions behind Patterson-based methods hold – $6n$-dimensional searches "overkill" the molecular replacement problem by unnecessarily doubling the dimensionality of the search space.

It should be mentioned, however, that this very property of ignoring evidence obtained from the Patterson function, makes these methods more robust and suitable for problems where the assumptions behind the Patterson-based methods are not satisfied.

3.   Implementation

A space group general computer program has been developed which implements the method described in Section 2 (program *Queen of Spades*, see Section 5.3 for availability details). As is always the case with Monte Carlo algorithms, the efficiency of the minimisation depends greatly on the optimal (or otherwise) choice of (i) an annealing schedule which specifies how the temperature of the system will vary with time, (ii) of the temperature (or temperature range) that will be used during the simulations, and (iii) of a suitable (for the problem under examination) target function whose value is to be optimised. The following paragraphs discuss these three points in more detail and present additional information about another issue that is important for the specific application, namely the bulk-solvent correction.

Annealing schedules: The current implementation of the program supports four annealing modes. In the first mode, the temperature is kept constant throughout the minimisation. The second is a slow-cooling mode, with the temperature linearly dependent on the simulation time. The third mode (which is the default) supports a logarithmic schedule for which the temperature $T(k)$ at each step k of the simulation is given by $T(k) = T_o/\log(k)$, where T_o is the starting temperature. In the last mode, the temperature of the system is automatically adjusted in such a way, as to keep the fraction of moves made against the gradient of the target function constant and equal to a user-defined value.

Temperature limit determination: It is possible to automatically obtain reasonable estimates of the temperature required for a constant and logarithmic temperature run, and of a temperature range for a slow-cooling run. This is achieved by monitoring the variation of the average value of the target function as a function of the temperature during a short slow-cooling simulation which is started from a sufficiently remote (high) temperature (this is similar to a specific heat plot from statistical mechanics, see [18]). It should be noted, however, that newer versions of the program come with pre-defined temperature limits when used in the automatic mode.

Target function selection: In other simulated annealing problems the target function (whose value is to be optimised) is an integral part of the problem and is, thus, not a matter of choice. In crystallographic problems, however, the issue of which function to optimise has been (and is some cases, still is) hotly debated. As was mentioned in Section 2, the currently distributed version of the program supports three user-selectable target functions: the conventional crystallographic R-factor and two correlation-based targets, the first of which is calculated over the amplitudes and the second over the intensities of the reflections. Still, the current thinking in the field clearly points the way to the theoretical (and, nowadays, practically achievable) superiority of a maximum-likelihood function (see contribution by Read, R., this volume, and references therein). The major problems with the implementation of a maximum-likelihood target in the context of the stochastic multidimensional search described in this communication are that (i) it is not clear how to estimate the σ_A curve [19] based on the necessarily small number of reflections used by this method, (ii) the σ_A curve would have to be recalculated at each and every step of the algorithm, and (iii) for most of the time, these calculations would be pointless, given that the majority of the sampled configurations during a minimisation are completely wrong (random) structures. It should also be noted that in contrast with the situation encountered with macromolecular refinement, stochastic molecular replacement is blessed with an extremely high ratio of observations to parameters (usually in the order of a few hundred reflections per parameter), and that the model is (by being the result of an independent structure determination) totally unbiased towards the observed data.

Bulk-solvent correction: The absence of a bulk-solvent correction from the molecular replacement calculations is a serious problem. Not only it introduces a systematic error for all data to approximately 6 or 5 Å resolution, but also, necessitates the application of a low-resolution cutoff (commonly at $\approx$15 Å) to compensate for the absence of a suitable correction. This low resolution cutoff, in turn, introduces series termination errors and further complicates the target function landscape, making the identification of the global minimum more difficult (see also [20]). Because at each and every step of the minimisation, we have a complete model for the target crystal structure, it is – at least in principle – possible to perform a proper correction for the presence of bulk solvent, as described, for example, by [21, 22]. The problem, of course, is that if at each step we had to calculate a mask for the protein component, followed by several rounds of refinement for the parameters of the solvent, the resulting program would be too slow to be practical. There is, however, a much faster but less accurate bulk solvent correction method (known as the exponential scaling model algorithm), which is based on Babinet's principle [23]. This correction can and has been implemented in the 'Queen of Spades' program (see [24] for more details).

4. Program parallelisation

The multimodel, multidimensional approach described does have its cost: with a few thousand unique reflections in a high symmetry space group and with more than two search models, a typical *Queen of Spades* run would take well over 2–3 weeks of CPU time on the fastest personal workstations. The way forward for such computationally intensive calculations is of course parallelisation. We have very recently [25] implemented (and released with

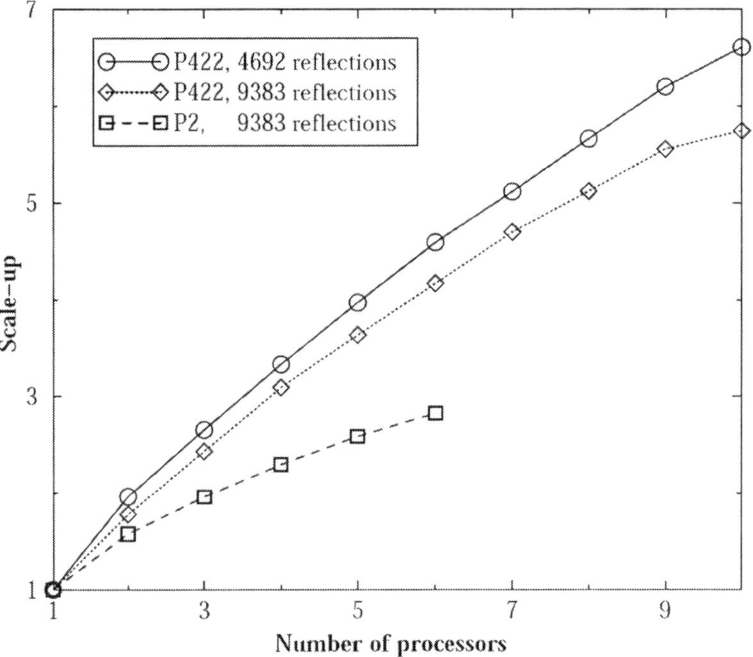

Figure 1. Parallel scale-up of *Queen of Spades* for three different problems.

the latest program distribution) a fine-grained parallelisation of the program using the message passing interface (MPI) paradigm. The parallel version will scale better as the number of unique reflections decreases and the symmetry increases. This is shown in Figure 1 which shows the parallel scale-up of the program for three different problems.

5.   Results

5.1.   APPLICATION TO A 17-DIMENSIONAL, THREE-MODEL PROBLEM

This example is based on the crystal structure of the complex of nuclear factor of activated T cells (NFAT), Fos, and Jun with DNA, corresponding to the Protein Data Bank (PDB) entry 1A02 (space group $P2_1$ with cell dimensions $a = 64.7$, $b = 85.6$, $c = 83.4$, $\beta = 112°$). To reduce the dimensionality of the problem, we have treated Fos and Jun as one (rigid) search model, leaving us with three models (NFAT, Fos-Jun, and the DNA molecule). We used real data deposited with the PDB (entry r1a02sf.ent), and to make the example more realistic we modified the deposited coordinates by subjecting (without experimental restraints) the starting models to energy minimisation for 500 cycles. The resulting models deviated from the deposited structures with root mean square (rms) deviations of 1.1, 1.5, and 2.2Å for the NFAT, Fos-Jun, and DNA respectively, and were used as search models in the ensuing molecular replacement calculations. The *Queen of Spades* run was performed in its fully automatic mode requiring only the PDB files containing the search models and a free-format ASCII file containing the observed data that the program should use for the calculation (all data between 19.5 and 4 Å resolution for this example). In this default mode, the program performed five independent minimisations each lasting 30 million Monte Carlo moves and taking approximately 73 h of CPU time on a personal computer equipped with a 1.8 GHz Pentium IV processor, 1 GB of random access memory and a proper operating system (GNU/Linux, RedHat distribution v.7.3). The total physical memory requirements of the program amounted to 196 MB (a significant proportion of which corresponds to high resolution volumes of the molecular transforms and, thus, does not have to be resident in memory). All minimisations used the strongest 70% of all reflections with $F/\sigma(F) > 2$ (4,943 reflections in total), with 10% of these being reserved for statistical cross-validation. The target function for the minimisation was $[1 - Corr(F_o, F_c)]$, where $Corr(F_o, F_c)$ is the linear correlation coefficient between the observed and calculated structure factor amplitudes. A Boltzmann annealing schedule was used with the temperature T at step k given by $T = T_o/\log(k)$, where T_o is the starting temperature for the minimisation (set to 0.070 [arbitrary units] in the default program mode).

Figure 2 shows the evolution of the average values of the target function (and its cross-validated counterpart) versus Monte Carlo moves for three

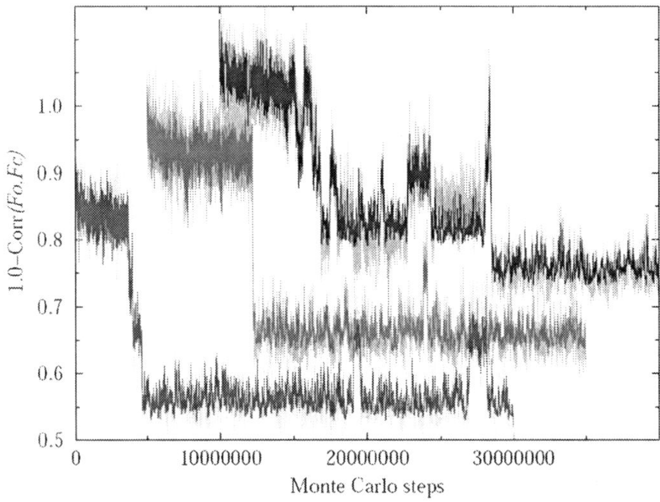

Figure 2. Evolution of the average values of the target function versus Monte Carlo moves. To minimise overlap successive graphs have been translated by 0.1 units along *y* and by 5 million moves along *x*. Each graph includes both the target function (foreground, dark colour) and its cross-validated value (background, light colour).

minimisations. All five minimisations resulted in closely related structures, with three of them (first, third, and fifth minimisations) converging to the correct crystal structure.

5.2. APPLICATION TO A 23-DIMENSIONAL, FOUR-BODY PROBLEM

The target crystal structure is the monoclinic form of the A31P mutant of Rop [26]. This form (space group $C2$ with cell dimensions $a = 94.4$, $b = 24.3$, $c = 64.5$, $\beta = 130°$) contains the equivalent of one complete 4-α-helical bundle

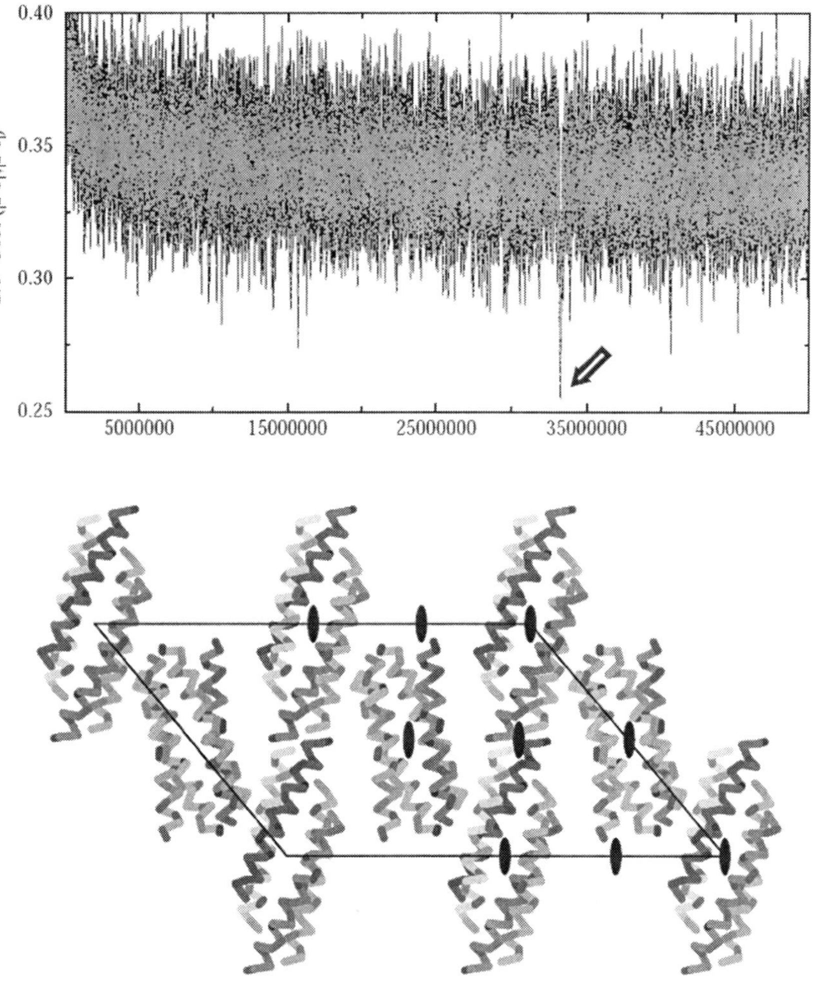

Figure 3. The top panel shows the evolution of the average value of the target function versus Monte Carlo moves. The arrow point indicates the position of the local minimum that led to elucidation of the structure. The lower panel shows a view of the corresponding crystal packing.

per asymmetric unit and only 35% solvent. For the successful run, we used the strongest 70% of all reflections between 15 and 3.5 Å resolution with $F/\sigma(F) > 2$ (999 reflections in total). Ten per cent of these reflections was reserved for statistical cross-validation. The target function for the minimisation was again the $[1 - Corr(F_o,F_c)]$ target. A Boltzmann annealing schedule was used with the temperature T at step k given by $T = T_o/\log(k)$, where T_o is the starting temperature for the minimisation (set to 0.070 [arbitrary units] for the successful minimisation). We performed five independent minimisations each lasting 50 million Monte Carlo moves and taking approximately 36 h of CPU time on a 800 MHz Intel Pentium III-based machine. The search model (four identical and independent copies) was a 26 residue long polyalanine helix (extracted from 1rpo.pdb, residues 4–29) and consisting of 129 atoms (less than 13% of the total number of atoms in the crystallographic asymmetric unit). Figure 3a shows the evolution of the average values of the target function versus Monte Carlo moves for the successful fifth minimisation and Figure 3b shows a view of the corresponding packing arrangement down the [010] axis. The search models are clearly organised as two independent 4-α-helical bundles centred on crystallographic twofold axes. Although the best solution from this method had one of the helices placed with the wrong polarity (and mistranslated by one helical turn in a direction parallel to the helical axis), it was, nevertheless, sufficiently close to the correct solution to allow structure determination to proceed to completion [27].

5.3. PROGRAM AVAILABILITY

Queen of Spades is free open-source software. The program's distribution includes source code, documentation, example scripts, and stand-alone executables (for the uniprocessor version) and is immediately available for download from the following world wide web address: http://www.mbg.duth.gr/~glykos/ or from the various mirrors generously provided by the Collaborative Computational Project number 14.

References

1. Sheriff, S., Klei, H.E., and Davis, M.E. (1999) Implementation of a six-dimensional search using the AMoRe translation function for difficult molecular-replacement problems. *Journal of Applied Crystallography*, **32**: 98–101.
2. Liu, Q., Weaver, A.J., Xiang, T., Thiel, D.J., and Hao, Q. (2003) Low-resolution molecular replacement using a six-dimensional search. *Acta Crystallographica*, **D59**: 1016–1019.
3. Strokopytov, B.V., Fedorov, A., Mahoney, N.M., Kessels, M., Drubin, D.G., and Almo, S.C. (2005) Phased translation function revisited: structure solution of the cofilin-homology domain from yeast actin-binding protein 1 using six-dimensional searches. *Acta Crystallographica*, **D61**: 285–293.

4. Jamrog, D.C., Zhang, Y., and Phillips, G.N. Jr. (2003) SOMoRe: a multi-dimensional search and optimization approach to molecular replacement. *Acta Crystallographica*, **D59**: 304–314.

5. Jiang, F. and Rao, Z. (2001) A new implementation of the molecular replacement method using a six-dimensional Patterson vector search. *Journal of Synchrotron Radiation*, **8**: 1051–1053.

6. Jogl, G., Tao, X., Xu, Y., and Tong, L. (2001) COMO: a program for combined molecular replacement. *Acta Crystallographica*, **D57**: 1127–1134.

7. Chang, G. and Lewis, M. (1997) Molecular replacement using genetic algorithms. *Acta Crystallographica*, **D53**: 279–289.

8. Kissinger, C.R., Gehlhaar, D.K., and Fogel, D.B. (1999) Rapid automated molecular replacement by evolutionary search. *Acta Crystallographica*, **D55**: 484–491.

9. Kissinger, C.R., Gehlhaar, D.K., Smith, B.A., and Bouzida, D. (2001) Molecular replacement by evolutionary search. *Acta Crystallographica*, **D57**: 1474–1479.

10. Glykos, N.M. and Kokkinidis, M. (2000) A stochastic approach to molecular replacement. *Acta Crystallographica*, **D56**: 169–174.

11. Glykos, N.M. and Kokkinidis, M. (2001) Multidimensional molecular replacement. *Acta Crystallographica*, **D57**: 1462–1473.

12. Glykos, N.M. and Kokkinidis, M. (2004) Molecular replacement with multiple different models. *Journal of Applied Crystallography*, **37**: 159–161.

13. McGreevy, R.L. and Pusztai, L. (1988) Reverse Monte Carlo simulation: a new technique for the determination of disordered structures. *Molecular Simulation*, **1**: 359–367.

14. Keen, D.A. and McGreevy, R.L. (1990) Structural modelling of glasses using reverse Monte Carlo simulation. *Nature*, **344**: 423–425.

15. Metropolis, N., Rosenbluth, A.W., Rosenbluth, M.N., Teller, A.H., and Teller, E. (1953) Equation of state calculations by fast computing machines. *Journal of Chemical Physics*, **21**: 1087–1092.

16. Ingber, L.A. (1993) Simulated annealing: practice versus theory. *Journal of Mathematical and Computational Modelling*, **18**: 29–57.

17. Navaza, J., Panepucci, E.H., and Martin, C. (1998) On the use of strong Patterson function signals in many-body molecular replacement. *Acta Crystallographica*, **D54**: 817–821.

18. Kirkpatrick, S., Gelatt, C.D., and Vecchi, M.P. (1983) Optimization by simulated annealing. *Science*, **220**: 671–680.

19. Read, R.J. (1997) Model phases: probabilities and bias. *Methods Enzymology*, **277**: 110–128.

20. Fokine, A. and Urzhumtsev, A. (2002) On the use of low-resolution data for translation search in molecular replacement. *Acta Crystallographica*, **A58**: 72–74.

21. Badger, J. (1997) Modelling and refinement of water molecules and disordered solvent. *Methods Enzymology*, **277**: 344–352.

22. Jiang, J.S. and Brunger, A.T. (1994) Protein hydration observed by X-ray diffraction: solvation properties of penicillopepsin and neuraminidase crystal structures. *Journal of Molecular Biology*, **243**: 100–115.

23. Moews, P.C. and Kretsinger, R.H. (1975) Refinement of the structure of carp muscle calcium-binding parvalbumin by model building and difference Fourier analysis. *Journal of Molecular Biology*, **91**: 201–228.

24. Glykos, N.M. and Kokkinidis, M. (2000) On the distribution of the bulk-solvent correction parameters. *Acta Crystallographica*, **D56**: 1070–1072.

25. Glykos, N.M. (2005) Qs v.1.3: a parallel version of Queen of Spades (submitted).

26. Glykos, N.M. and Kokkinidis, M. (2003) Structure determination of a small protein through a 23-dimensional molecular replacement search. *Acta Crystallographica*, **D59**: 709–718.

27. Glykos, N.M. and Kokkinidis, M. (2004) Structural polymorphism of a marginally stable 4-α-helical bundle. Images of a trapped molten globule? *Proteins*, **56**: 420–425.

LIKELIHOOD-BASED MOLECULAR REPLACEMENT IN *PHASER*

RANDY J. READ, AIRLIE J. MCCOY, AND
LAURENT C. STORONI

*Department of Haematology, University of Cambridge,
Cambridge Institute for Medical Research, Wellcome
Trust/MRC Building, Hills Road, Cambridge CB2 0XY, UK*

Abstract: Likelihood is being applied to an ever-expanding range of problems in macromolecular crystallography, from experimental phasing through to map calculation and structure refinement. We have shown that likelihood also provides a better measure for rotation and translation searches in the molecular replacement method, but searches using the full likelihood functions are computationally expensive. In our new program *Phaser* we have implemented fast approximations to the likelihood targets, which allow rotation and translation searches to be carried out quickly, but maintaining the sensitivity of the likelihood searches. *Phaser* contains a number of other innovations, including a likelihood-based correction for anisotropy, and a pruned tree search for multiple molecules, which allows it to solve many difficult molecular replacement problems automatically.

Keywords: molecular replacement; maximum likelihood; automation.

1. Introduction: likelihood in crystallography

When confronted by an optimization problem, most scientists turn automatically to least squares. In fact, this is the choice that would usually be made if the optimization problem were approached from a maximum likelihood perspective. The idea behind maximum likelihood is that the best model is most consistent with the data; consistency, or likelihood, is measured by the probability that the data would have been measured given the model. Maximizing a likelihood target for observations with Gaussian measurement errors turns out to be equivalent to minimizing a least squares target. Gaussian measurement errors are almost ubiquitous because of the central limit theorem, which says that as a number of random variables (e.g., sources of error) are added together, the resulting sum has a probability distribution

R.J. Read and J.L. Sussman (eds.): Evolving Methods for Macromolecular Crystallography, 91–100
© 2007. *Springer*

that tends towards a Gaussian. Most experiments have a number of possible sources of error, so most observations have errors that can be modeled well with Gaussian distributions.

However, there are reasons why the least squares target is often not the correct choice in crystallography, and it is often worth stepping back to first principles and applying maximum likelihood. The main reason is the phase problem. In many problems in crystallography, particularly for macromolecules, measurement errors are only a minor source of error in predicting the data. Most of the error originates in differences between pairs of structure factors (e.g., calculated and observed, native, and derivative). Thanks to the central limit theorem, differences between structure factors also tend to have Gaussian distributions, but these Gaussian distributions apply to the structure factors with phases. The observations are amplitudes or intensities and, after integration to remove the unknown phases, their distributions are no longer Gaussian.

Because of this, likelihood has been playing an increasing role in crystallography. One of the earlier uses was in estimating σ_A parameters for phase probability distributions [1, 2], which can then be used for phase combination or to compute map coefficients with reduced model bias [2, 3]. The likelihood function used to estimate σ_A values is based on the probability of the observed structure factor amplitude given the calculated structure factor. Essentially the same likelihood function can be applied to structure refinement, optimizing the parameters of the atomic model from which structure factors are calculated [4–6]. Likelihood has also had a great impact on experimental phasing from isomorphous derivatives and anomalous scatterers. Likelihood functions for heavy-atom refinement and phasing [7, 8] have been implemented in the program SHARP [9]. As discussed elsewhere in these proceedings by Airlie McCoy, we have implemented likelihood targets for phasing by SAD and MIRAS in the experimental phasing modules of our program *Phaser*.

The use of likelihood as a target for molecular replacement was initially proposed by Bricogne [10], in a paper that described some approximations that could be adapted for fast Fourier transform (FFT)-based computations. We have built on that initial suggestion, adding a new rotation likelihood function, new fast approximations for rotation and translation searches, and allowing for multiple models through the application of multivariate structure factor statistics. All these are implemented in the molecular replacement modules of our program *Phaser*.

2. Structure factor probabilities

Crystallographic phasing methods all turn out to involve essential steps where related structure factors are compared: model with observed (as in molecular replacement), native with derivative, or derivative with calculated heavy-atom contribution. Information about one structure factor comes from its correlations with another structure factor. This information is statistical, so that it is necessary to consider the underlying probability distributions. As outlined in previous publications (reviewed in Read [3]), the correlation between two structure factors will depend on the size of the common substructure (how many atoms the structures share) and the size of the coordinate differences between the substructures.

We are often interested in the conditional distribution of one structure factor (such as the true $\mathbf{F}$) given another (such as $\mathbf{F}_C$ computed from a molecular replacement model). A proportion (D) of $\mathbf{F}_C$ will be correlated to $\mathbf{F}$, so that $D\mathbf{F}_C$ is the expected value of $\mathbf{F}$ given $\mathbf{F}_C$. D will tend to decrease with resolution, as coordinate differences become larger relative to the Bragg spacing. For differences between structure factors that are the sums of large numbers of atomic contributions, the central limit theorem will apply and we can assume that the distribution of $\mathbf{F}$ is a complex Gaussian or, equivalently, a symmetric two-dimensional (2D) Gaussian in the complex plane for an acentric structure factor, or a one-dimensional (1D) Gaussian for a centric structure factor [11].

The likelihood function is the probability distribution for the observations, which are intensities or structure factor amplitudes in crystallography. Therefore, the distribution for the phased structure factor has to be converted to a distribution for an amplitude or intensity. This is achieved by integrating over all possible values of the unknown phase angle. The resulting probability distribution (in the acentric case) is known as the Rice distribution in statistical literature, but also occurs frequently in crystallographic literature (e.g., Sim [12]). Because it does not take the form of a Gaussian, the use of least squares targets is not justified by the principle of maximum likelihood.

3. Molecular replacement likelihood functions

Traditionally, molecular replacement has been carried out with a divide-and-conquer algorithm, in which, first the orientation of the molecule is determined with a rotation function, then its position with a translation function. This is the approach we have taken in *Phaser*. It is also possible to carry out full $6n$-dimensional searches using either systematic [13] or stochastic [14, 15] algorithms.

3.1. TRANSLATION LIKELIHOOD FUNCTION

From a rotated and translated model we can compute a set of calculated structure factors, and it is appropriate to use the same likelihood function that is used in other applications with calculated structure factors, i.e., to estimate σ_A values for map coefficients [2] or to refine atomic models (e.g., reference [4]). This likelihood function would also be suitable for algorithms carrying out full $6n$-dimensional searches.

3.2. ROTATION LIKELIHOOD FUNCTION

In some cases, particularly rotation searches, there is an extra source of ambiguity that the likelihood function must account for. When the orientation of a molecule is known, but not its position, we can calculate the amplitudes of the molecular transform contributions from the symmetry-related copies of the model, but until we know its position we do not know the relative phases of these symmetry-related contributions. The addition of structure factor contributions with unknown relative phase is a random walk problem, reminiscent of the situation when the atoms comprising a crystal are known, but not their positions. As Wilson [16] showed, if there are sufficient atoms of similar scattering power in the unit cell, the distribution of structure factors can be approximated well as a Gaussian, from which the unknown phase (or sign) can be integrated out. In the case of molecular replacement searches, the number of molecular transform contributions is relatively small, calling the Gaussian approximation into question. On the other hand, molecular replacement problems only become difficult when there are many molecules in the unit cell or when the models are poor or incomplete, in which case the Gaussian approximation will be better justified. A slightly better approximation can be made by following a suggestion from Shmueli et al. [17] regarding crystals with heterogeneous compositions. The single biggest contribution to the structure factor can be taken as a partial structure factor (as for the translation search), with the rest of the contributions adding to the Gaussian noise. This is the approximation used in the rotation likelihood function in *Phaser*.

Apart from pure rotation searches, there are other circumstances where one must account for relative phase ambiguity. One important case is when the orientation and position of one or more molecules is known prior to the search for the orientation of another molecule. *Phaser* can deal with such cases, and when information about a partial solution is accounted for, the signal-to-noise of a rotation search is often increased significantly.

3.3. CALIBRATING THE LIKELIHOOD FUNCTIONS

The likelihood functions contain two parameters that depend on the completeness and accuracy of the model: the factor D giving the proportion of the calculated structure factor that is correlated to the true structure factor, and the variance of the estimate. In fact, both of these parameters can be determined from a single parameter, σ_A, which combines the effects of model completeness and accuracy. In applications to map calculation [2] and model refinement [4], σ_A is estimated as a function of resolution by optimizing a likelihood target. This would be too expensive for each step of a molecular replacement search, so instead we use an educated guess. The variation of σ_A with resolution can be predicted reasonably well using an equation that depends on the completeness of the model (usually known at the outset of a molecular replacement search) and the estimated root mean square (RMS) error of each component of the model. We have found that an equation derived by Chothia and Lesk [18], correlating sequence identity with main-chain RMS deviation, provides a useful prediction of the RMS error of molecular replacement models.

4. Fast likelihood-enhanced targets

Experience with the program *Beast* [19], which carries out brute-force rotation and translation likelihood calculations, demonstrated the increased sensitivity of likelihood. However, the increased sensitivity comes at a high computational cost, as the brute-force calculations take hours or even days, and there was clearly a need for faster methods.

Bricogne [10, 20] first suggested that appropriate approximations of the likelihood function could be used to carry out fast molecular replacement searches with likelihood-based methods. We have adopted the spirit of that suggestion, deriving Taylor series approximations to the rotation and translation likelihood functions, which can be computed rapidly by FFT methods. To combine the increased speed of the approximations with the optimal sensitivity of the full likelihood score, we select peaks from the fast searches and rescore them with the full likelihood targets, in a strategy similar to that used in *AMoRe* [21].

4.1. LIKELIHOOD-ENHANCED FAST ROTATION FUNCTION

Most conventional Patterson-based rotation searches are now carried out with variants of the fast rotation function described by Crowther [22], in which the Patterson is decomposed into spherical harmonics and Bessel functions so that the overlap integrals can be computed in Eulerian β sections by FFT. The same technique can be used to compute approximations of the likelihood function. The first-order term in a Taylor series approximation of the rotation likelihood function has the same functional form as equations given by Navaza [23], so the same techniques can be used to compute a likelihood-enhanced fast rotation function [24]. Compared to a Patterson-based fast rotation function, the likelihood-based version has resolution-dependent weights reflecting expected model accuracy, and terms accounting for the effect of previously known fixed contributions to the structure.

4.2. LIKELIHOOD-ENHANCED FAST TRANSLATION FUNCTION

Similarly, first- and second-order Taylor series approximations of the translation likelihood function can be evaluated by FFT methods [25], using algorithms devised by Navaza and Vernoslova [26] to compute fast correlation functions. The first-order approximation (termed LETF1 in *Phaser*), which is evaluated using only a single FFT, is used by default. In test calculations, this fast translation function has proven to be more sensitive than the fast correlation search, which requires three FFTs in its evaluation and consumes more memory.

5. Automating molecular replacement in *Phaser*

Part of *Phaser*'s success in solving a wide variety of difficult molecular problems is owed to the sensitivity of the likelihood targets, compared to the conventional Patterson-based targets. But much of its success is also due to automation features, in which *Phaser* keeps track of all plausible partial solutions and tries them in turn in a pruned tree search.

A typical automated *Phaser* run proceeds as follows:

- Correct input data for anisotropy, using an algorithm that adjusts anisotropic B-factors to make the data agree with a Wilson [16] distribution for isotropic data
- Carry out fast rotation search
- Rescore plausible fast rotation peaks with full rotation likelihood target
- Carry out fast translation search for each plausible orientation
- Rescore plausible fast translation peaks with full translation likelihood target
- Check plausible solutions for crystal packing
- Carry out rigid body refinement of solutions that pack successfully
- Prune duplicate solutions

If there is more than one molecule in the asymmetric unit, this process is repeated, fixing each plausible refined solution in turn and starting with the fast rotation search for the next molecule. If there is any ambiguity in the space group, *Phaser* can also test a list of possible space groups in the translation search for the first molecule. If there is more than one possible choice of model or ensemble of models, these can be tested in turn as well.

5.1. β-LACTAMASE–BLIP COMPLEX

The structure of a complex between β-lactamase and the β-lactamase inhibitor protein (BLIP) was originally solved with great difficulty by molecular replacement [27]. Using *Phaser*, this structure can be solved easily, and the process of structure solution illustrates the use of several features in *Phaser*. A default run in *Phaser* proceeds as follows:

- Correct input data for anisotropy, which is pronounced in this data set. At the limit of resolution (3 Å), the anisotropy corrections to the amplitudes differ by a factor of about 2.5 between the weakest and strongest directions. Without the anistropy correction, the structure solution is much more difficult.

- Carry out likelihood-enhanced fast rotation search for β-lactamase orientation, then rescore top orientations with the full rotation likelihood function. This results in one unambiguous peak, with a Z-score (number of RMS deviations above the mean score) of 12.2.

- Carry out likelihood-enhanced fast translation search for β-lactamase in the orientation from the rotation search, then rescore top translation vectors with full likelihood score. This results again in a single unambiguous peak, with a Z-score of 24.1.

- Check β-lactamase solution for packing, then carry out rigid body refinement of this solution.

- Fix β-lactamase solution while carrying out rotation search for BLIP component (fast search followed by rescoring). This results in two similar orientations, each with a Z-score of 5.7.

- Continue to fix β-lactamase solution while carrying out translation searches for each BLIP orientation from the previous step. This results in two similar solutions, the better of which has a Z-score of 30.3.

- Check solutions for packing, carry out rigid body refinement, then check whether solutions are equivalent. A single unique solution is obtained.

If *Phaser* is used to search for BLIP without fixing the β-lactamase solution, the signal is much weaker; although the correct orientation is still at the top of the list after the rotation search, the Z-score is only 4.1, and *Phaser* retains 15 trial orientations. If, in addition, the anisotropy correction is omitted, *Phaser* is unable to locate the BLIP component by itself, as the correct orientation does not appear in the list from the rotation search.

6. Future prospects

The molecular replacement algorithms in *Phaser* work well, but there are a number of potential improvements. Judging from the structures that *Phaser* still fails to solve, the most important is to take account of non-crystallographic symmetry (NCS). As Navaza et al. [28] discuss, we often have prior knowledge from native Patterson maps and self-rotation functions. The impact of translational NCS, in which more than one molecule is found in the same orientation, is the most serious. With translational NCS, the molecules related by translation have essentially the same molecular transforms, but they add up in phase or out of phase depending on the relative positions of the molecules. As a result, there is a systematic variation in observed intensity through the diffraction pattern, determined more by the NCS translation vector than by the structures of the molecules themselves. As long as the likelihood function fails to take account of this variation, such structures will be difficult to solve. In principle, we understand how to deal with this problem, but some work is required to implement the changes to the likelihood function and the refinement of the relevant parameters.

Useful information could also be obtained from rotational NCS, in which different copies of a molecule take on different orientations. It is often possible to determine the rotations that relate pairs of such molecules from a self-rotation function, and the information from this could be used to filter the output from the rotation function, leaving only sets of orientations that satisfy the known NCS.

7. Availability

Phaser is part of the Phenix package for automated crystallography [29] and will be distributed as part of the CCP4 [30] package. It is also available from our website, at http://www-structmed.cimr.cam.ac.uk/phaser.

Acknowledgements

We are grateful to Michael James and Natalie Strynadka for supplying the data for the β-lactamase complex test case. The research underlying *Phaser* is supported by a Wellcome Trust Principal Research Fellowship to RJR, and its development within the Phenix package is supported by NIH/NIGMS under grant No. 1P01GM063210. Anne M. Baker was supported by CCP4 to develop the original ccp4i interface, with assistance from Peter Briggs.

References

1. Lunin, V.Y. and Urzhumtsev, A.G. (1984) Improvement of protein phases by coarse model modification. *Acta Crystallographica*, **A40**: 269–277.
2. Read, R.J. (1986) Improved Fourier coefficients for maps using phases from partial structures with errors. *Acta Crystallographica*, **A42**:140–149.
3. Read, R.J. (1997) Model phases: probabilities and bias. *Methods in Enzymology*, **277**: 110–128.
4. Pannu, N.S. and Read, R.J. (1996) Improved structure refinement through maximum likelihood. *Acta Crystallographica*, **A52**: 659–668.
5. Bricogne, G. and Irwin, J. (1996) Maximum likelihood structure refinement: theory and implementation within BUSTER + TNT, in: *Macromolecular Refinement: Proceedings of the CCP4 Study Weekend*. Edited by Dodson, E., Moore, M., Ralph, A., and Bailey, S. Warrington: Daresbury Laboratory, pp. 85–92.
6. Murshudov, G.N., Vagin, A.A., and Dodson, E.J. (1997) Refinement of macromolecular structures by the maximum-likelihood method. *Acta Crystallographica*, **D53**: 240–255.
7. Bricogne, G. (1991) A maximum-likelihood theory of heavy-atom parameter refinement in the isomorphous replacement method, in: *Isomorphous Replacement and Anomalous Scattering: Proceedings of the CCP4 Study Weekend*. Edited by Evans, P.R. and Leslie, A.G.W. Warrington: Daresbury Laboratory, pp. 60–68.
8. Read, R.J. (1991) Dealing with imperfect isomorphism in multiple isomorphous replacement, in: *Crystallographic Computing 5: From Chemistry to Biology*. Edited by Moras, D., Podjarny, A.D, and Thierry, J.C. Oxford: Oxford University Press, pp. 158–167.
9. La Fortelle, E. de. and Bricogne, G. (1997) Maximum-likelihood heavy-atom parameter refinement for multiple isomorphous replacement and multiwavelength anomalous diffraction methods. *Methods in Enzymology*, **276**: 472–494.
10. Bricogne, G. (1992) A statistical formulation of the molecular replacement and molecular averaging methods, in: *Molecular Replacement: Proceedings of the CCP4 Study Weekend*. Edited by Wolf, W., Dodson, E.J., and Glover, S. Warrington: Daresbury Laboratory, pp. 62–75.
11. Read, R.J. (1990) Structure-factor probabilities for related structures. *Acta Crystallographica*, **A46**: 900–912.
12. Sim, G.A. (1959) The distribution of phase angles for structures containing heavy atoms. II. A modification of the normal heavy-atom method for non-centrosymmetrical structures. *Acta Crystallographica*, **12**: 813–815.
13. Sheriff, S., Klei, H.E., and Davis, M.E. (1999) Implementation of a six-dimensional search using the *AMoRe* translation function for difficult molecular replacement problems. *Journal of Applied Crystallography*, **32**: 98–101.
14. Kissinger, C.R., Gehlhaar, D.K., and Fogel, D.B, (1999) Rapid automated molecular replacement by evolutionary search. *Acta Crystallographica*, **D55**: 484–491.
15. Glykos, N.M. and Kokkinidis, M. (2000) A stochastic approach to molecular replacement. *Acta Crystallographica*, **D56**: 169–174.
16. Wilson, A.J.C. (1949) The probability distribution of X-ray intensities. *Acta Crystallographica*, **2**: 318–321.
17. Shmueli, U., Weiss, G.H., Kiefer, J.E., and Wilson, A.J.C. (1984) Exact random-walk models in crystallographic statistics. I. Space groups P1-bar and P1. *Acta Crystallographica*, **A40**: 651–660.
18. Chothia, C. and Lesk, A.M. (1986) The relation between the divergence of sequence and structure in proteins. *The EMBO journal*, **5**: 823–826.
19. Read, R.J. (2001) Pushing the boundaries of molecular replacement with maximum likelihood. *Acta Crystallographica*, **D57**: 1373–1382.
20. Bricogne, G. (1997) Bayesian statistical viewpoint on structure determination: basic concepts and examples. *Methods Enzymol*, **276**: 361–423.
21. Navaza, J. (1994) *AMoRe*: an automated package for molecular replacement. *Acta Crystallographica*, **A50**: 157–163.

22. Crowther, R.A. (1972) The fast rotation function, in: *The Molecular Replacement Method*. Edited by Rossmann, M.G. New York: Gordon and Breach, pp. 173–178.
23. Navaza, J. (2001) Implementation of molecular replacement in *AMoRe*. *Acta Crystallographica*, **D57**: 1367–1372.
24. Storoni, L.C., McCoy, A.J., and Read, R.J. (2004) Likelihood-enhanced fast rotation functions. *Acta Crystallographica*, **D60**: 432–438.
25. McCoy, A.J., Grosse-Kunstleve, R.W., Storoni, L.C., and Read, R.J. (2005) Likelihood-enhanced fast translation functions. *Acta Crystallographica*, **D61**: 458–464.
26. Navaza, J. and Vernoslova, E. (1995) On the fast translation functions for molecular replacement. *Acta Crystallographica*, **D51**: 445–449.
27. Strynadka, N.C., Jensen, S.E., Alzari, P.M., and James, M.N. (1996) A potent new mode of beta-lactamase inhibition revealed by the 1.7A X-ray crystallographic structure of the TEM-1-BLIP complex. *Nature Structural Biology*, **3**: 290–297.
28. Navaza, J., Panepucci, E.H., and Martin, C. (1998) On the use of strong Patterson function signals in many-body molecular replacement. *Acta Crystallographica*, **D54**: 817–821.
29. Adams, P.D., Grosse-Kunstleve, R.W., Hung, L.-W., Ioerger, T.R., McCoy, A.J., Moriarty, N.W., Read, R.J., Sacchettini, J.C., Sauter, N.K., and Terwilliger, T.C. (2002) PHENIX: developing new software for automated crystallographic structure determination. *Acta Crystallographica*, **D58**: 1948–1954.
30. Collaborative Computational Project, Number 4 (1994) The CCP4 suite: programs for protein crystallography. *Acta Crystallographica*, **D50**: 760–763.

AUTOMATED STRUCTURE DETERMINATION WITH PHENIX

PAUL D. ADAMS, PAVEL V. AFONINE,
RALF W. GROSSE-KUNSTLEVE, NIGEL W. MORIARTY,
NICHOLAS K. SAUTER, AND PETER H. ZWART

*Lawrence Berkeley National Laboratory, One Cyclotron
Road, BLDG 64R0121, Berkeley, CA 94720, USA*

KRESHNA GOPAL, THOMAS R. IOERGER,
LALJI KANBI, ERIK MCKEE, AND REETAL K. PAI

*Department of Computer Science, Texas A&M University,
301 H.R. Bright Building, 3112 TAMU, College Station, TX
77843, USA*

LI-WEI HUNG AND THIRU RADHAKANNAN

*Biophysics Group, Mail Stop D454, Los Alamos National
Laboratory, Los Alamos, NM 87545, USA*

AIRLIE J. MCCOY, RANDY J. READ, AND
LAURENT C. STORONI

*Department of Haematology, University of Cambridge,
Cambridge Institute for Medical Research, Wellcome
Trust/MRC Building, Hills Road, Cambridge, CB2 0XY, UK*

TOD D. ROMO AND JAMES C. SACCHETTINI

*Department of Biochemistry and Biophysics, Texas A&M
University, 103 Biochemistry/Biophysics Building, 2128
TAMU, College Station, TX 77843, USA*

THOMAS C. TERWILLIGER

*Los Alamos National Laboratory, Mailstop M888, Los
Alamos, NM 87545, USA*

R.J. Read and J.L. Sussman (eds.): Evolving Methods for Macromolecular Crystallography, 101–109
© 2007. *U.S. Government*

Abstract: A new software system called PHENIX (Python-based Hierarchical ENvironment for Integrated Xtallography) has been developed for the automation of crystallographic structure solution. This provides algorithms to go from reduced intensity data to a refined molecular model, and facilitates structure solution for both the novice and expert crystallographer. Here, we review the major features of PHENIX, including the different user interfaces, and briefly describe the recent advances in infrastructure and algorithms.

Keywords: PHENIX; automation; X-ray crystallography.

1. Introduction

There is a pressing need for high-throughput structure determination to support efforts such as structural genomics [1, 2] and structure-based development of therapeutics. In this high-throughput mode structures need to be solved significantly faster than has routinely been achievable in the last few years. This high-throughput structure determination requires automation to reduce the obstacles related to human intervention. Manual interpretation of complex numerical data has a significant subjective component [3] that can lead to delays in reaching the final structure. Thus, the automation of structure solution is essential as it has the potential to produce minimally biased models more efficiently. We are developing the PHENIX software to address these needs [4, 5].

2. PHENIX design and implementation

2.1. HYBRID PROGRAMMING

The core PHENIX infrastructure is based on a tight integration between reusable software components written both in a compiled language and a flexible scripting language. Prior experience implementing the Crystallography and NMR System [6] has shown that this promotes highly efficient software development. High-level algorithms such as complex refinement protocols or phasing procedures are most rapidly developed in a scripting language. By contrast, numerically intensive core algorithms such as the computation of structure factors or discrete Fourier transforms must be implemented in a compiled language for performance reasons.

PHENIX uses Python (http://python.org/) as the scripting language and C++ as the compiled language; see Grosse-Kunstleve et al. [7] and

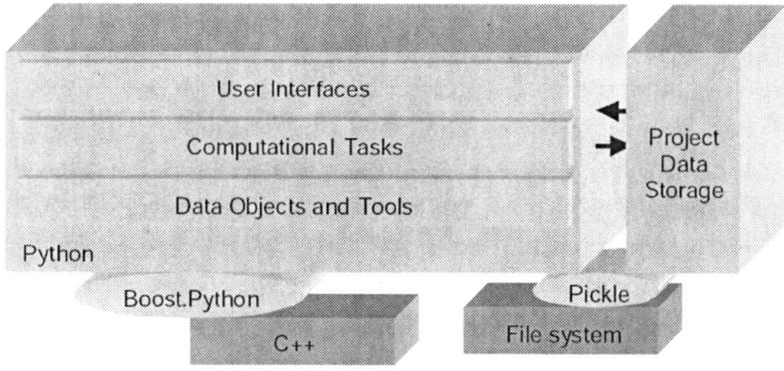

Figure 1. The architecture of the PHENIX system.

Grosse-Kunstleve and Adams [8] for further details about these choices. This relies on the Boost.Python Library [9] for conveniently integrating C++ and Python. It is used to directly connect C++ classes and functions to Python without obscuring the C++ interface. Most components of the PHENIX software are tested on the commonly available computing platforms: Redhat Linux, HP Tru64, SGI Irix version 6.5, and Windows 2000. The Macintosh OS-X platform will also be supported in the future as the necessary tools become available.

2.2. PHENIX ARCHITECTURE

PHENIX builds upon Python, the Boost.Python Library, and C++ to provide an environment for automation and scientific computing (Figure 1). Many of the fundamental crystallographic building blocks, such as data objects and tools for their manipulation, are provided by the Computational Crystallography Toolbox (*cctbx* [7]). The computational tasks which perform complex crystallographic calculations are then built on top of this. Finally, there are a number of different user interfaces available in order to use PHENIX. In order to facilitate automated operation there is the Project Data Storage (PDS) that is used to store and track the results of calculations.

3. Crystallographic algorithms

A number of algorithms for structure solution have been implemented. A new substructure searching procedure, called HySS (hybrid substructure search), makes use of Patterson and direct methods to locate anomalously scattering or heavy atoms for experimental phasing [10]. This algorithm incorporates criteria that automatically determine when a correct solution is likely to have been found. The SOLVE program is used for experimental maximum likelihood phasing for MAD/SAD and MIR/SIR data [11]. New algorithms for phasing are also being developed for use in PHENIX [12, 13]. Initial phases can also be obtained using molecular replacement incorporating maximum likelihood targets [14–16]. The use of these new targets increases the success rate of this method using search models of lower structural similarity. The phases obtained from experimental phasing or molecular replacement are optimized by the application of maximum likelihood density modification algorithms, implemented in the RESOLVE program, to produce minimally biased electron density maps [17, 18]. Electron density maps are automatically interpreted using template matching [18–20] as implemented in RESOLVE, and pattern recognition methods as implemented in TEXTAL [21, 22]. The automated map interpretation algorithms are iterated with maximum likelihood refinement targets [23, 24] and simulated annealing optimization algorithms [25–27]. With this combination of tools automated structure solution is possible even when diffraction data are available only to modest resolution limits (2.5–3 Å) (see Figure 2).

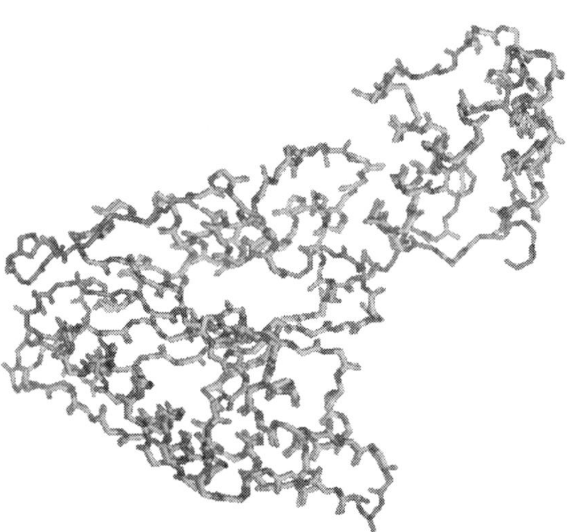

Figure 2. Model automatically built from a single Se-SAD wavelength at 2.4 Å resolution using the AutoSol and AutoBuild wizards (main chain only shown). Approximately 75% of the backbone and 70% of the side chains are built without manual intervention within 4 h. Figure created with PyMOL [28].

4. Phenix user interfaces

From the inception of PHENIX it was clear that different user interfaces would be required depending on the needs of a diverse user community. There are currently three different user interfaces, each described in Sections 4.1–4.3.

4.1. COMMAND-LINE

For some situations a very simple command-line interface is most effective. This is particularly so when rapid results are required, such as real-time structure solution at the synchrotron beamline. We have developed tools that facilitate ease of use at the early stages of structure solution, data analysis (iotbx.reflection_statistics), and anomalous substructure location (phenix.hyss). For example, the user interface for the data statistics command is:

```
iotbx.reflection_statistics < reflection-file >
[<reflection-file>...]
```

The format of the reflection file is automatically determined and the data interpreted appropriately. For each data set overall and bin-wise statistics are reported for completeness, anomalous signal (if anomalous data are present), and intensity moments. The highest non-origin peaks from a native Patterson synthesis are reported to help detect translational non-crystallographic symmetry. The possible twinning operations for the unit cell and symmetry are also automatically calculated and listed.

4.2. STRATEGIES

The PHENIX strategy interface provides a way to construct complex networks of tasks to perform a higher-level function (Figure 3). For example, the steps required to go from initial data to a first electron density map in a SAD experiment can be broken down into well-defined tasks, which can be reused in other procedures. Instead of requiring the user to run these tasks in the correct order they are connected together by the software developer, and can thus be run in an automated way. However, because the connection between tasks is dynamic they can be reconfigured or modified, and new tasks introduced as necessary if problems occur. This provides the flexibility of user input and control, while still permitting complete automation when decision-making algorithms are incorporated into the environment. The tasks and their connection into strategies rely on the use of plain text task files written using the Python scripting language. This enables the computational algorithms to be used easily in a non-graphical environment.

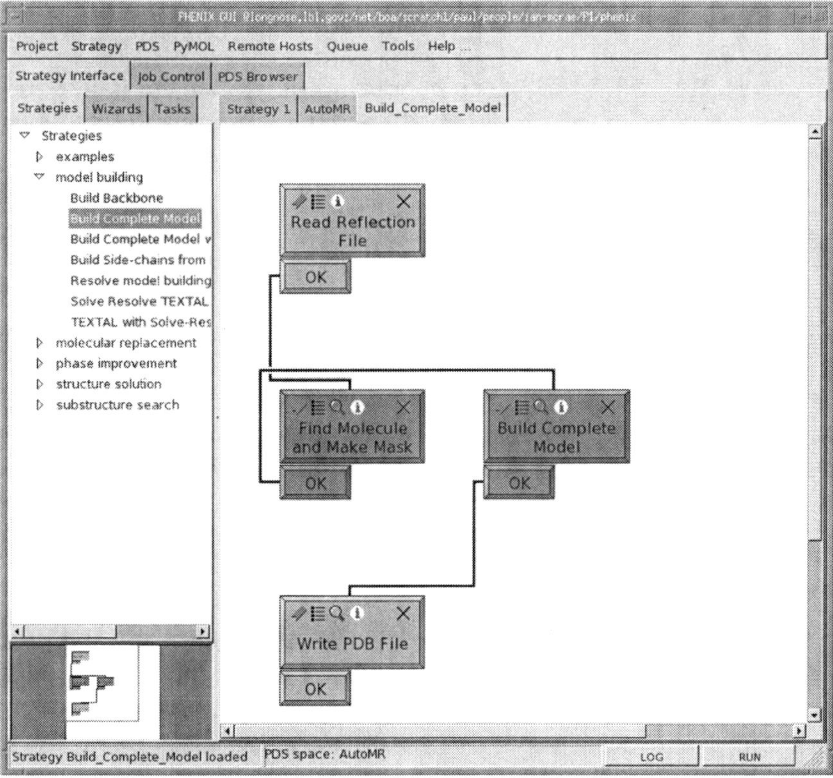

Figure 3. Example of the PHENIX strategy interface, showing a TEXTAL strategy for automated model building.

The PHENIX GUI permits strategies to be visualized and manipulated. These manipulations include loading a strategy distributed with PHENIX, customizing and saving it for future recall.

Current strategies available include:

- Substructure location using HySS

- Structure solution (MAD/SAD and SIR(AS)/MIR(AS)) combining HySS, SOLVE, and RESOLVE

- Structure solution and initial refinement using HySS, SOLVE/RESOLVE, TEXTAL, and phenix.refine

- Automated maximum likelihood molecular replacement using PHASER

- Phase improvement using maximum likelihood methods in RESOLVE

- Model building using TEXTAL or RESOLVE

4.3. WIZARDS

The decision making in strategies is local, with decisions being made at the end of each task to determine the next route in the path. This is analogous with how crystallographers typically make decisions during structure solution; a program is run, the outputs manually inspected and a decision made about the next step in the process. By contrast, a wizard provides a user interface that can make more global decisions, by considering all of the available information at each step in the process. As the name suggests, wizards are designed to lead users through a complex process, making automatic decisions when possible, but prompting the user for additional information when necessary. The wizard interface uses the same graphical environment as the strategies, but consists of only a single input or output area (Figure 4).

Currently, there are the following wizards available:

- AutoSol: automated structure solution (from processed data to a first map and model) that uses HySS, SOLVE and RESOLVE

- AutoMR: automated molecular replacement, using PHASER

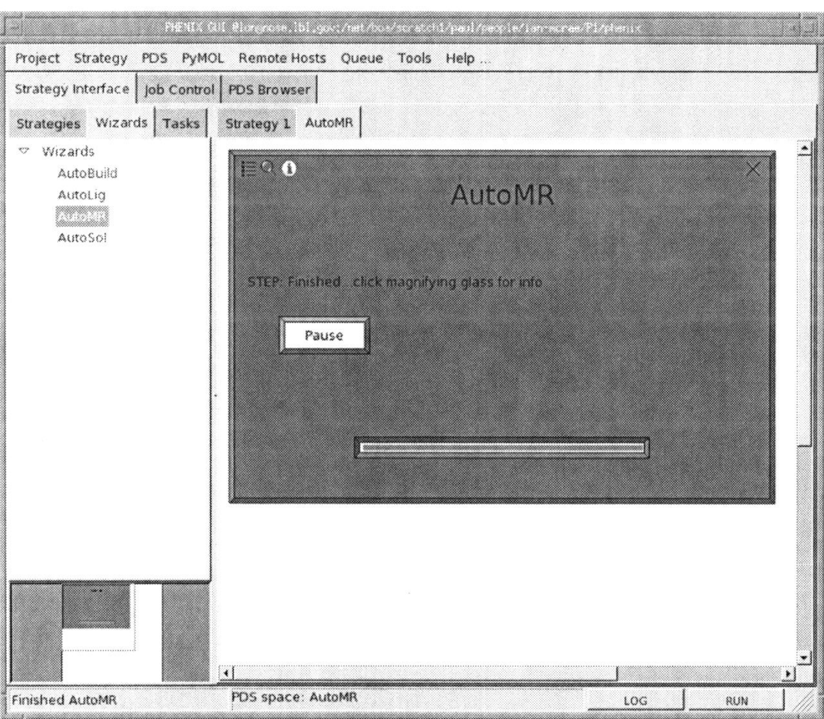

Figure 4. Example of the PHENIX wizard interface, with the AutoMR wizard shown.

- AutoBuild: automated model building or rebuilding combined with structure refinement that uses RESOLVE and phenix.refine
- AutoLig: automated ligand fitting into difference density maps using RESOLVE once a close to complete protein model is available

5. Conclusions

The development of PHENIX is a collaborative project whose primary goal is the creation of a comprehensive, integrated system for automated crystallographic structure determination. The PHENIX infrastructure is also designed to be open and easily shared with other researchers. The use of the Python scripting language facilitates interfacing with the system and its use in different contexts. Projects have already begun that use PHENIX as the basis for new tools for single particle electron cryo-microscopy image reconstruction, and refinement of macromolecular structures using neutron diffraction data.

Acknowledgements

This work was funded by NIH/NIGMS under grant number 1P01GM063210, with initial funding to PDA from the Department of Energy under contract No. DE-AC03-76SF00098.

References

1. Burley, S.K., Almo, S.C., Bonanno, J.B., Capel, M., Chance, M.R., Gaasterland, T., Lin, D., Sali, A., Studier, F.W., and Swaminathan, S. (1999) *Nature Genetics*, **23**: 151–157.
2. Sali, A. (1998) *Nature Structural and Molecular Biology*, **5**: 1029–1032.
3. Mowbray, S.L., Helgstrand, C., Sigrell, J.A., Cameron, A.D., and Jones, T.A. (1999) *Acta Crystallographica*, **D55**: 1309–1319.
4. Adams, P.D., Grosse-Kunstleve, R.W., Hung, L.-W., Ioerger, T.R., McCoy, A.J., Moriarty, N.W., Read, R.J., Sacchettini, J.C., Sauter, N.K. and Terwilliger, T.C. (2002) *Acta Crystallographica*, **D58**: 1948–1954.
5. Adams, P.D., Gopal, K., Grosse-Kunstleve, R.W., Hung, L.-W., Ioerger, T.R., McCoy, A.J., Moriarty, N.W., Pai, R.K., Read, R.J., Romo, T.D., Sacchettini, J.C., Sauter, N.K., Storoni, L.C., and Terwilliger, T.C. (2004) *Journal of Synchrotron Radiation*, **11**: 53–55.
6. Brunger, A.T., Adams, P.D., Clore, G.M., Gros, P., Grosse-Kunstleve, R.W., Jiang, J.-S., Kuszewski, J., Nilges, N., Pannu, N.S., Read, R.J., Rice, L.M., Simonson, T., and Warren, G.L. (1998) *Acta Crystallographica*, **D54**: 905–921.
7. Grosse-Kunstleve, R.W., Sauter, N.K., Moriarty, N.W., and Adams, P.D. (2002) *Journal of Applied Crystallography*, **35**: 126–136.
8. Grosse-Kunstleve, R.W. and Adams, P.D. (2003a) *IUCr Computing Commission Newsletter*, **1**.
9. Abrahams, D. and Grosse-Kunstleve, R.W. (2003) *C/C++ Users Journal*, **21**: 29–36.

10. Grosse-Kunstleve, R.W. and Adams, P.D. (2003b) *Acta Crystallographica*, **D59**: 1966–1973.
11. Terwilliger, T.C. and Berendzen, J. (1999) *Acta Crystallographica*, **D55**: 849–861.
12. Read, R.J. (2003) *Acta Crystallographica*, **D59**: 1891–1902.
13. McCoy, A.J., Storoni, L.C., and Read, R.J. (2004) *Acta Crystallographica*, **D60**: 1220–1228.
14. Read, R.J. (2001) *Acta Crystallographica*, **D57**: 1373–1382.
15. Storoni, L.C., McCoy, A.J., and Read, R.J. (2004) *Acta Crystallographica*, **D60**: 432–438.
16. McCoy, A.J., Grosse-Kunstleve, R.W., Storoni, L.C., and Read, R.J. (2005) *Acta Crystallographica*, **D61**: 458–464.
17. Terwilliger, T.C. (2002) *Acta Crystallographica*, **D58**: 2082–2086.
18. Terwilliger, T.C. (2001) *Acta Crystallographica*, **D57**: 1755–1762.
19. Terwilliger, T.C. (2003a) *Acta Crystallographica*, **D59**: 38–44.
20. Terwilliger, T.C. (2003b) *Acta Crystallographica*, **D59**: 45–49.
21. Ioerger, T.R. and Sacchettini, J.C. (2002) *Acta Crystallographica*, **D58**: 2043–2054.
22. Holton, T., Ioerger, T.R., Christopher, J.A., and Sacchettini, J.C. (2000) *Acta Crystallographica*, **D56**: 722–734.
23. Lunin, V.Y., Afonine, P.V., and Urzhumtsev, A.G. (2002) *Acta Crystallographica*, **A58**: 270–282.
24. Pannu, N.S., Murshudov, G.M., Dodson, E.J., and Read, R.J. (1998) *Acta Crystallographica*, **D54**: 1285–1294.
25. Brunger, A.T., Kuriyan, J., and Karplus, M. (1987) *Science*, **235**: 458–460.
26. Adams, P.D., Pannu, N.S., Read, R.J., and Brunger, A.T. (1997) *Proceedings of the National Academy of Sciences of the USA*, **94**: 5018–5023.
27. Adams, P.D., Pannu, N.S., Read, R.J., and Brunger, A.T. (1999) *Acta Crystallographica*, **D55**: 181–190.
28. DeLano, W.L. (2002) The PyMOL Molecular Graphics System (http://www.pymol.org).

DENSITY MODIFICATION IN MAIN

DUŠAN TURK

Jozef Stefan Institute, Department of Biochemistry and Molecular Biology, Jamova 39, 1000 Ljubljana, Slovenia

Abstract: Electron density modification methods are an indispensable part of any *de novo* macromolecular crystal structure determination and can be of crucial importance for determining structures solved by the molecular replacement method. In MAIN, a number of density modification procedures have been implemented. They encompass tools like solvent flattening and electron density averaging. In addition, maps can be generated using maximum likelihood weighting as well as by the "kick map" approach. Kick maps have been shown to be a good alternative to maximum-likelihood maps, when model bias has to be revealed. These approaches are intergrated with the MAIN model building and refinement tools, which also allow multicrystal averaging and refinement with non-crystallographic constraints across a variety of crystal forms.

Keywords: density modification; density averaging; map calculation; kick maps.

1. Introduction

Density modification is understood as a procedure in which density of solvent as well as molecular regions are (iteratively) modified. Historically, electron density averaging routines were developed immediately after the model building tools and solvent picking procedures in the early 1990s as a part of my Ph.D. thesis [1]. Cathepsin B was the first structure where averaging with MAIN has been applied [2]. The other features such as map calculation, solvent flattening and refinement, model validation, and autobuilding tools have been gradually appended later.

The phase problem must be solved before MAIN can become useful. This implies that the initial density map should be either constructed from contributions of heavy atoms or calculated from an appropriately positioned molecular model. In the cases of large structures containing a variety of identical subunits with known position, internal symmetry, and approximate size, phases can be derived by density averaging within MAIN too.

111

R.J. Read and J.L. Sussman (eds.): Evolving Methods for Macromolecular Crystallography, 111–122
© 2007. *Springer*

Input to MAIN are molecular models, space group and unit cell data, structure factors, and density maps. From these data density maps can be calculated and manipulated. Molecular models, structure factors, and density maps are interoperable forms of data within the algebra of the MAIN program.

From the perspective of a user density modification procedures are triggered within MAIN by clicking the four menu items (RMS_FIT, MAK_MASK, DM_PREP, DM_NEXT). The macros underlying them are configured with the help of the "main_config" tools. In the user perspective, the solvent flattening procedure differs from density averaging in the "RMS_FIT" part, the use of which is confined to the non-crystallographic symmetry (NCS) tools, density averaging, and model building.

Maps are input to density modification procedures; therefore, their calculation is described first. Density averaging is considered as a special case of solvent flattening, therefore, the latter is described before the density averaging.

2.   Map calculations

Fo, phase-combined and difference maps (2Fo–Fc, Fo–Fc) can be computed using least squares and maximum-likelihood weighting schemes and averaged kick map and kick omit maps schemes. The maximum likelihood functions for map calculations and refinement entered MAIN with the help of Urzhumtzev and Afonine in 2002 [3]. Reciprocal space anisotropic B-factor and bulk-solvent correction terms based on contribution of solvent derived from the molecular envelope [4] are also included.

The map calculations can be activated as a part of the MAP_CALC menu or configured by the density modification scripts in the case of density modification procedures.

Phase combination is performed using Hendrickson—Lattman coefficients and structure factors from the model. The structure factors from the model are calculated in MAIN followed by phase combination in the external "Sigma A" program [5].

The MAP_CALC menu items BULK_SOL, ANISO_B, ML_MAPS, RE_PHASE, PHAS_CMB, FOBS_MAP, 2FO_FC_M, FO_FC_MA cover map calculations apart from kick maps. The BULK_SOL, ANISO_B, ML_MAPS are flags, which turn on and off the corresponding options.

Among these, only kick maps and kick omit maps are unique to MAIN, therefore they are briefly described below.

2.1. KICK MAPS

The procedure that randomly displaces atoms from their original positions is called "kicking". The concept of kick maps was introduced in MAIN 97

(ACA meeting abstracts, 1997) and used the first time to verify the orientation of an octapeptide attached to the cathepsin H surface [6]. Kicking can be used also in refinement.

Kicking atoms results in their random displacements along X, Y, and Z coordinates

```
do i = 1, natoms
   X(i)_mod = X(i)_orig + KICK * 2 * (rnd - 0.5))
   Y(i)_mod = Y(i)_orig + KICK * 2 * (rnd - 0.5))
   Z(i)_mod = Z(i)_orig + KICK * 2 * (rnd - 0.5))
end do
```

It is expected that the structure factors calculated from kicked atoms contain less model bias than the factors calculated from models coming directly out of a refinement cycle, during which atoms interact with each other via chemical and crystallographic energy terms. With the increasing maximum size of the kick, the resulting molecular models are increasingly further away from the original model and thereby become increasingly less related to other atoms in the model (Figure 1). The coordinates of kicked models do contain large errors. Therefore, the resulting maps are averaged. An averaged kick map is calculated from a series of models, where each of them was generated with a different random number seed (Figure 1f). (In practice, it means each time a model is to be kicked the starting seed is increased by one.) It can be demonstrated that in an averaged kick map calculated from a series of difference maps with maximum sizes of the kick up to 1.4 Å, a substantial part of noise has been filtered out (Figure 1e). The resulting averaged kicked maps are namely clearer and reveal more molecular features than the standard non-kicked least squares-weighted electron density maps.

The kick maps are comparable to maximum likelihood maps and more often than not they differ from the maximum likelihood maps in the direction which is closer to the "true" solution of the structure.

A homologue of the kicking is a molecular dynamics run at high temperature. The differences are

- Kick acts instantaneously and after displacement atoms are returned back to their initial positions
- Atomic model can be distorted to the level where the structure has lost the correct "chemical" sense and becomes unmanageable for procedures using the chemical description of a model

When a part of the model is omitted from the structure factor calculation the resulting maps are termed "kick omit maps". Intriguingly, although the kick omit maps can be calculated in a window through the complete

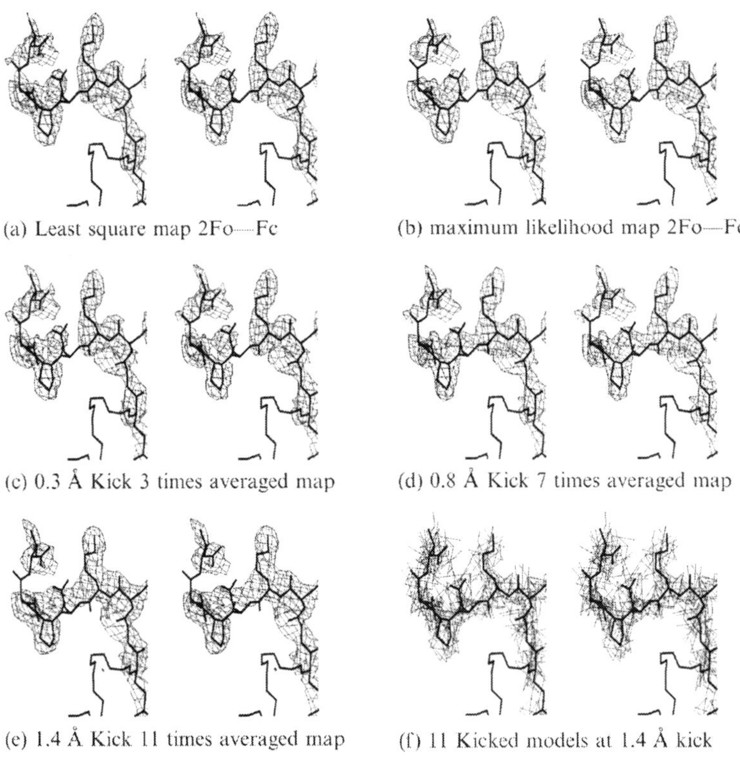

(a) Least square map 2Fo—Fc

(b) maximum likelihood map 2Fo—Fc

(c) 0.3 Å Kick 3 times averaged map

(d) 0.8 Å Kick 7 times averaged map

(e) 1.4 Å Kick 11 times averaged map

(f) 11 Kicked models at 1.4 Å kick

Figure 1. Electron density maps at 3 Å resolution calculated with different approaches are showed in figures from (a) to (e), whereas in the figure (f), eleven kicked models of the region are shown with a *thin line*. On the background of all figures the non-perturbed region of the atomic model of amodytoxin A is showed with a *thick line*. Only the region from Glu 12 to Ser 20 is showed to demonstrate obvious differences between the approaches used in map calculations. The gap between Lys15 and Asn16 closes only with rather large kicks of 0.8 and 1.4 Å.

sequence of a model, they are generally not better than the kick maps themselves, suggesting that kicking itself deals quite well with the model bias.

Since the kick-map calculation takes a short amount of time (repetition of structure factor calculation form) it can be applied very often during an interactive session and the resulting density can help a user to build through dubious places.

Users are in control of the maximum size of the kick and the number of cycles used in averaging. In MAIN 2000, the kick-map calculations are triggered and controlled with the items (KICK_0.3, KICK_??, KICK_SEE, MAP_+_??, OMIT_MAP, KICK_MAP) of the "CALC_MAP" menu.

The choice of the maximal kick size depends on the resolution of diffraction data used in the calculation. On the lower side of the kick the values below one tenth of resolution do not unbias the map, whereas on the high end, errors still cancel out by kicks larger than 1.5 Å in medium to lower resolution range. With 3.0 Å data a maximum kick size of 0.8 Å, seven times averaged seems reasonable (Figure 1d), whereas for calculation of a map at 2.0 Å only kicks up to 0.5 Å and up to five times averaged are useful.

3. Solvent flattening

3.1. SEPARATION OF MOLECULAR FROM SOLVENT REGIONS

As the first step, statistical approaches are used to separate solvent from protein regions based on real space procedures of Wang [7] and Abrahams [8] and their FFT analogues [9]. Before the decision about the solvent content is finalized, it makes sense to visually inspect the score map at the suggested cut-off for possible manual corrections with atomic models. Score maps can also be calculated independently by clicking the "SCORE_MAP" item and visualized at an appropriate cut-off.

When a partial atomic model is available, it is used independently from the envelopes proposed by the statistical methods. Model derived envelopes can replace the statistical envelopes as the model becomes complete. Models can be generated to mask the space only and do not have to contribute to structure factor calculation. A poly Phe helix can mask a substantial region, thereby enabling a user to gain control over the envelope assignment. Alternatively, skeleton atoms can be used too.

3.2. MOLECULAR PART MANIPULATION

The density within the part marked as the molecular region can be scaled, shifted, or otherwise manipulated. For example, the large fluctuations of density at a heavy atom site can be damped to a reasonable height or simply set to 0. The underlying density can also be combined with density coming from a partial molecular model in a similar way as performed by Bhat and Blow [10].

Alternatively, one may also use the atomic model to approximate the density in the molecular region.

3.3. USING ATOMIC MODEL

The best approximation of density of a molecular model is the model itself. Evidently, solution of the phase problem does not, in every case, provide an electron density map which allows building of complete molecular models.

When looking at a "kicked" atomic model it became obvious that in spite of the large errors and chemically erroneous structures, kicked sets of atomic coordinates still represent the "true" structure. This indicates that also models with large local coordinate errors correspond to the "true" structure. Therefore, in MAIN atomic models built from true amino acid residues and combined with polycarbon models derived from a skeleton are used. Skeleton-derived atoms were also used in the program PRISM [11], whereas hybrid models are also successfully used by the ARP/wARP program [12]. At resolutions below 2.5–2.7 Å such atomic models cannot be reasonably refined. One can, however, utilize such models as part of the density modification procedures. The advantage of such models is that they are essentially complete, so that scaling of calculated structure factors to the observed data becomes more accurate and the absences of atoms and errors in their positions more random than systematic.

3.4. SOLVENT PART MANIPULATION

The solvent region can be flattened [7], flipped [8], shifted, or scaled.

3.5. TRANSFORMATION THROUGH RECIPROCAL SPACE

The real power of solvent flattening procedures lies in Fourier transformation of a density map into reciprocal space. During the transformation additional structure factors can be added to the set previously included. The obtained structure factors can be combined with experimental phases [5] or used for an Fo or a difference 2Fo–Fc map calculation, which then serves as input to the next cycle.

4. Electron density map averaging

The simplest way of increasing the accuracy of a measurement is to average observations. The larger the number of observations, the more accurate is the average. Quite often asymmetric units of crystals of macromolecules contain multiple copies of identical subunits allowing averaging of density. Positions of these identical subunits are not related by crystal symmetry, therefore such packings are termed NCS. Sometimes also the alternative terms local symmetry and non-crystallographic similarity are used. The term local symmetry emphasizes the contrast to crystallographic symmetry which spreads throughout the crystal, whereas the NCS is confined to the asymmetric unit. The term non-crystallographic similarity is used to emphasize that such arrangements are quite often not symmetrical at all.

There are two types of non-crystallographic symmetry: proper (also called spherical) and improper. Molecules of an asymmetric unit are related by proper symmetry when they can be superimposed on each other by rotation(s) about a centre of rotation only, whereas in the case of improper symmetry, operators of superposition include also translational components besides the rotational. Therefore, procedures for proper and improper symmetry averaging differ. In proper symmetry averaging, equivalent areas can be separated, although it is generally better that they are not, whereas in the cases of improper symmetery averaging separate masks for each subunit must be built. Occasionally proper and improper symmetries are combined and can be layered. Density can be averaged also between a variety of crystal forms.

To perform real space electron density averaging some initial set of phases, molecular envelopes (also termed masks), and superposition operators (rotation and translation parameters) are required.

4.1. MOLECULAR ENVELOPE CREATION

For averaging, molecular envelopes can be derived from atomic models, with the initial atomic radii starting at a maximal value of 6 Å. At the molecular interfaces the atomic radii are reduced to half of the distance to the closest atom from the surrounding molecule related by crystallographic or non-crystallographic symmetry. As the model becomes more complete, the sizes of atomic radii are reduced.

When no molecular model is available, then the envelope can be created from the density map itself. First, skeleton atoms are derived from the density map. Then the skeleton atoms are displayed and edited manually using a three-dimensional (3D) GUI interface. In this process parts belonging to separate molecules are identified. Check for symmetry overlaps is included. After parts of the skeleton are assigned, the skeleton atoms can be used for mask derivation in the same way as molecular model atoms.

4.2. DERIVING DENSITY SUPERPOSITION PARAMETERS

The operators between equivalent parts can be easily constructed by superposition of molecular models. It is easy to envisage that also a heavy atom-phased case simplifies to a molecular model case as soon as parts of a model (or positions of heavy atoms) can be assigned to local subunits. From the model, parameters are derived by an root mean square (RMS) fitting procedure. MAIN uses segment identifiers (segment names) to assign local subunits of the model from which superposition parameters are derived. They need to be assigned before the RMS fitting procedures can be applied.

When this is not possible (no molecular model can be built and heavy atom positions do not reveal any non-crystallographic symmetry), the superposition operators have to be derived from the density map itself. In this process exploitation of the data provided by the self-rotation function of the Patterson map may prove useful.

Phases should, however, be good enough to reveal the molecular envelopes, which then in MAIN need to be marked (using interactive map skeletonization and assignment). The density within the envelopes is transferred to a large unit cell to enable a molecular replacement program such as AMoRe [13] to find the density superposition parameters. These parameters are then refined by using real space density grid point superposition.

As soon as the improved density allows building the first fragments of structure, the model is used for calculation of superposition parameters. Each time after the molecular model is expanded, it should be refined against the background map using NCS constraints.

In the case of proper symmetry, superposition parameters have less freedom. When the positions of molecules are related by an n-fold rotational axis, the n-fold of the axis should be fixed. In the case of icosahedral symmetry, however, all superposition matrices are fixed.

4.3. MOLECULAR PART MANIPULATION

Density is averaged at each masked grid point. The density from an equivalent point is obtained by linear interpolation from the surrounding eight grid points using the 8-4-2-1 point interpolation.

The averaged density within each molecular mask is scaled, and it can also be shifted or otherwise manipulated. Afterwards the masked density points fill the unit cell using the crystal symmetry operators.

Density in the regions of space allocated to molecules, but not present in multiple copies, is used to fill the unit cell using crystal symmetry operators too.

4.4. THE REST OF THE CYCLE

The subsequent solvent part manipulation and transformation of the resulting density are the same as in the solvent flattening procedure.

4.5. WHEN NOT TO CYCLE

When the resulting density map after the Fourier transform looks worse than after averaging, cycling through reciprocal space should be postponed.

Instead, molecular masks or superposition parameters, or both should be improved before continuing.

The Fo–Fc map and Patterson maps can be averaged, however, their Fourier transforms do not make sense. The Fo–Fc difference maps may substantially improve the density corresponding to a ligand or solvent molecules, whereas averaging of the Patterson map can help to refine the rotation axis and angle and reveal the number of subunits present as in the case of the proteasome [14], where electron microscopy data and the unmodified Patterson could not clearly differentiate between the six fold and seven fold rotational axis.

5. User interface

5.1. CONFIGURABLE STRATEGIES

MAIN config tools generate scripts, which can deal with any number of molecules and the complete variety of solvent flattening and map calculation possibilities provided by MAIN, including the use of the external program SIGMAA [5] for phase combination.

Organization of molecules into groups based on segment names:

```
-g |--group) defines NCS groups: specify the number of
             groups followed by each group of segments
             embraced in " "
             1 EACH | MOLA MOLB
             2 EACH | IA IB
-s |--strategy) defines  averaging  scripts  strategy -
             specify parameters for each group sep-
             arately [1 EACH]
             EACH: average each molecule separately
             (default)
             ONE: average one and distribute the
             averaged density to others
             LINK: use operators from another group
```

Density modification menu:

```
[cyclic/c]   cyclic density modification cycles (use
             FFT recalculation of structure factors
             from a map) 0 stops before re_fft [1]
[extend/e)   extend phase in range [ - ]
[phases/p]   the first phases - from model or file
             [MODEL]
[input/i]    input map kind [2FOFC]
```

```
[output/o]    output map kind [2FOFC]
[atom/a]      mask atom maximal radius [6.0]
[solv/s]      solvent flattening method [WANG]
              solvent flattening sphere radius [10.0]
              solvent content [0.5]
[flip/f]      flip of solvent region: 0.0
[hist/h]      B-value of atom used for histogram
              matching: 0.0
[mask_dir/m]  directory for mask files:
[rot_tran/rt] directory for rotation and transla-
              tional macros:
[show/sh]     show current MY_MAIN settings
[go]          go to "com" and "cmds" macros genera-
              tion steps

enter your choice - i

map for input to dm [3fo2fc/2fofc/sigma_a/fofc/fobs/
          comb/copy] - current map is [2FOFC]=
```

5.2. User written extensions

There are cases that are not covered by the "main_config" tools. For these editing of MAIN macros and knowledge of MAIN syntax are required.

The procedures based on improper symmetry for density-averaging parts related by proper non-crystallographic symmetry can usually be used, however, there are cases, which do not allow splitting of molecular envelopes to subunits due to their overlaps. In such cases, molecular envelopes must encompass all molecules constituting such a group.

Similarly, density superposition operators may require adjustments. (At least the RMS fitting procedure should have quite often a fixed rotation in the polar angle system, whereas orientation and position of the rotational axes should be left unconstrained.)

Use of proper and improper averaging procedures combined in layers is not yet supported by the "main_config" tools. Such macros need to be written by a user himself.

In MAIN documentation, only the case with two crystal averaging is provided [15] and should be edited. The MAIN command language allows averaging electron density of any number of crystal forms [16] in any combination and layers.

Patterson map averaging is not supported by the menu items.

By mastering MAIN algebra also Patterson map editing, masking the envelopes, and back transforming is possible, for example, to assist in search of heavy atom positions as done by Brandstetter in the case of the tricorn protease structure [17].

6. Concluding remarks

Within MAIN automated and manual tools for density modifications are integrated with the tools for model building, map calculations, structure refinement, and validation. Interactive interfaces enable a user to choose between more or less automated approaches and specify the order in which they are applied, thereby leaving the user the control over the progress of macromolecular crystal structure determination and enabling the most efficient use of the software.

References

1. Turk, D. (1992) Weiterentwicklung eines Programms fuer Molekuelgraphik und Elektrondichte-Manipulation und seine Anwendung auf verschiedene Protein-Strukturaufklärungen. Ph.D. thesis, Technische Universität, München. (The only page of the thesis in German is its title page, all others are in English.)
2. Musil, D., Zucič, D., Turk, D., Engh, R.A., Mayr, I., Huber, R., Popovič, T., Turk, V., Towatari, T., Katunuma, N., and Bode, W. (1991) The refined 2.15-A X-ray crystal structure of human liver cathepsin B: the structural basis for its specificity, The *EMBO Journal*, **10**: 2321–2330.
3. Lunin, V.Y., Afonine, P.V., Urzhumtsev, A.G. (2002) Likelihood-based refinement. I. Irremovable model errors. *Acta Crystallographica*, **A58**: 270–282.
4. Fokine, A. and Urzhumtsev, A. (2002) Flat bulk-solvent model: obtaining optimal parameters. *Acta Crystallographica: Section D Biological Crystallography*, **58**, 1387–1392.
5. Read, R.J. (1986) Improved Fouier coefficients for maps using phases from partial structures with errors. *Acta Crystallographica*, **A42**: 140–149.
6. Gunčar, G. Podobnik, M., Pungerčar, J., Štrukelj, B., Turk, V., and Turk, D. (1998) Crystal structure of porcine cathepsin H determined at 2.1Å resolution: location of the mini-chain C-terminal carboxyl group defines cathepsin H aminopeptidase function. *Structure*, **6**, 51–61.
7. Wang, B.C. (1985) Resolution of phase ambiguity in macromolecular crystallography. In *Diffraction methods in for Biological Macromolecules*, Edited by Wyckof H., New York. Academic press, vol. 115. pp. 90–112.
8. Abrahams, J.P. and Leslie, A.G.W. (1996) Methods used in the structure determination of bovine mitochondrial F1 ATPase. *Acta Crystallographica*, **D52**: 30–42.
9. Leslie, A.G.W. (1987) A reciprocal-space method for Calculating a molecular envelope using the algorithm of B.C. Wang. *Acta Crystallographica*, **A43**: 134–136.
10. Bhat, T.N. and Blow, D.M. (1983) A method for refinement of partially interpreted protein structures including a procedure for scaling between a model and an electron-density map. *Acta Crystallographica*, **A39**: 166–170.
11. Wilson, C. and Agard, D.A. (1993) PRISM: automated crystallographic phase refinement by iterative skeletonization. *Acta Crystallographica*, **A49**: 97–104.
12. Perrakis, A., Morris, R., and Lamzin, V.S. (1999) Automated protein model building combined with iterative structure refinement. *Nature Structural Biology*, **6**: 458–463.
13. Navaza, J. (1994) AMoRe: an automated package for molecular replacement. *Acta Crystallographica*, **A50**: 157–163.

14. Löwe, J., Stock, D., Jap, B., Zwickl, P., Baumeister, W., and Huber, R. (1995) *Science* **268**: 533–539.
15. Turk, D., Podobnik, M., Kuhelj, R., Dolinar, M., and Turk, V. (1996) Crystal structures of human pro-cathepsin B at 3.2 and 3.3 Å resolution reveal an interaction motif between a papain like cysteine protease and its propetide. *FEBS Letters*, **384**: 211–214.
16. Baumann, U., Wu, S., Flaherty, K.M., and McKay, D.B. (1993) Three-dimensional structure of the alkaline protease of Pseudomonas aeruginosa: a two-domain protein with a calcium binding parallel beta roll motif. *The EMBO Journal*, **12**: 3357–3364.
17. Brandstetter, H., Kim J.S., Groll, M., and Huber, R. (2001) Crystal structure of the tricorn protease reveals a protein disassembly line. *Nature* **414** (6862): 466–470.

AB INITIO PHASING STARTING FROM LOW RESOLUTION

VLADIMIR LUNIN AND NATALIA LUNINA

*Institute of Mathematical Problems of Biology RAS,
Pushchino, 142290 Russia*

ALEXANDRE URZHUMTSEV

*Université Henry Poincaré, Nancy I, 54506,
Vandoeuvre-lès-Nancy, France
Current address:1G BCM, 67404 Illkirch and IBMC,
67084 Strasbourg, France*

Abstract: Low-resolution phasing is important in the study of large macromolecular complexes and in the case of crystals of limited diffraction power. It allows defining molecular positions in the unit cell, molecular envelopes, and, in favorable cases, secondary structure elements. A multifiltering phasing method is discussed that is designed for the use of different weak low-resolution criteria of a phase set quality.

Keywords: phase problem; *ab initio* phasing; direct phasing; low resolution; connectivity; likelihood; Fourier syntheses; histograms; glob models; macromolecules.

1. Introduction

1.1. *AB INITIO* (DIRECT) PHASING

Phase determination is a necessary step to transform a set of diffraction magnitudes into images of the electron density distribution. Conventional ways to solve this problem in macromolecular crystallography involve either additional diffraction experiments (using modified wavelengths or crystal content) or knowledge of the model of a homologous object. There exist also methods capable to solve the structure using a single set of structure factor magnitudes and some general properties of electron density distribution (the "atomicity" as a rule). In this paper, we call these phasing methods *ab initio* or "direct" although sometimes these terms are reserved for a more broad

R.J. Read and J.L. Sussman (eds.): Evolving Methods for Macromolecular Crystallography, 123–133
© 2007. *Springer*

meaning. Such methods are routine in "small molecules" crystallography and last decade they came into macromolecular field (for a possible review see [21]), however, their application requires a high-resolution data set (about 1 Å, usually) that is not always possible in up-to-date macromolecular crystallography. In this paper, we consider an opposite case, namely low-resolution phasing, when the low-resolution edge reflections only are measured [6, 14, 15, 17]. The phasing of such reflections cannot provide one with a fine structure of the studied object, but nevertheless the information obtained may play a significant role for a further success in the structure determination.

We suppose below that the input of the phasing procedure is the structure factor magnitudes $\{F^{obs}(\mathbf{s})\}, \mathbf{s} \in S$ for a set S of reciprocal-space vectors and some additional "general type" information on the object under study. The goal of the phasing is to find structure factors phases $\{\varphi(\mathbf{s})\}, \mathbf{s} \in S$ that allow the Fourier synthesis of the electron density be calculated

$$\rho(\mathbf{r}) = \frac{1}{V_{cell}} \sum_{s \in S} F^{obs}(\mathbf{s}) \exp\left[i\varphi(\mathbf{s})\right] \exp\left[-2\pi i(\mathbf{s}, \mathbf{r})\right].$$

1.2. LOW-RESOLUTION PHASING

In this paper, we use the term "low resolution" to note several dozens (or a few hundreds) reflection of the lowest resolution for the given crystal. Depending on the size of the unit cell a Fourier synthesis calculated with these structure factors may present different information on the object studied. If this restricts the resolution approximately by $d_{min} > 16$ Å, the information that can be extracted from Fourier syntheses concerns mostly the macromolecular position in the unit cell and its envelope (Figures 1 and 2). This information can simplify the translation and eventually rotation search in the molecular replacement and facilitates the use of complementary sources of information like electron microscopy reconstructed images. If the resolution exceeds approximately 8 Å, Fourier syntheses may show α-helixes (Figure 3) and at the resolution of about 4 Å β-sheets become visible. Syntheses of an intermediate resolution 16 Å $> d_{min} > 8$ Å are the most difficult for interpretation and overcoming this resolution interval presents the largest difficulties in *ab initio* phasing.

The low-resolution reflections are often ignored in the process of structure solution mainly by the following reasons:

- Experimental difficulties when collecting low-resolution diffraction data, especially for crystals with very large unit cells

- Strong influence of the bulk solvent on low-resolution structure factors that makes it difficult to use the corresponding magnitudes in conventional phasing methods

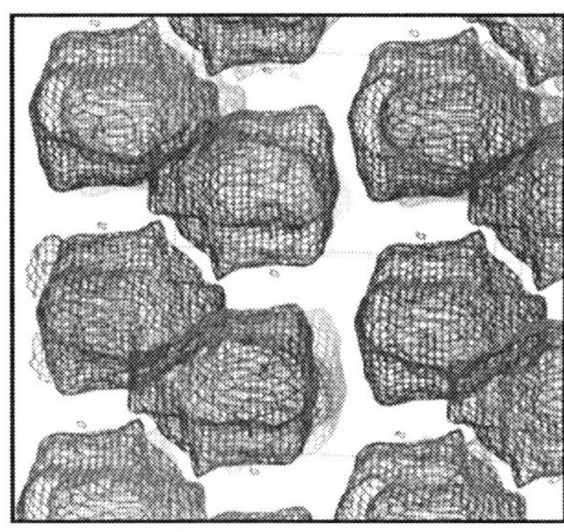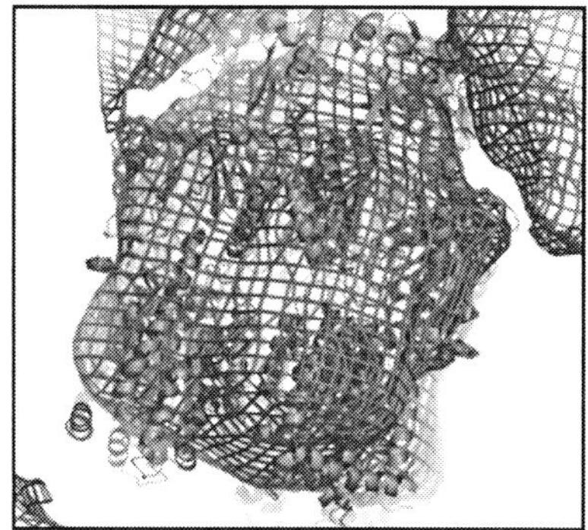

Figure 1. Ab initio phased Fourier synthesis for membrane protein AcrB (40 Å resolution, 29 reflections) [18]. Packing of trimer envelopes and the published structure 1iwg [19] manually superposed with the envelope.

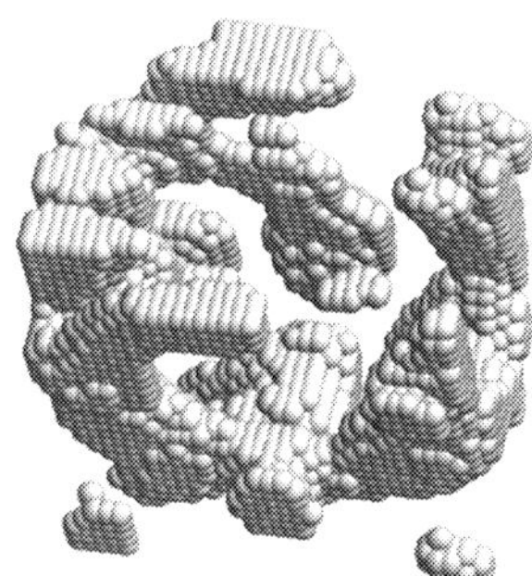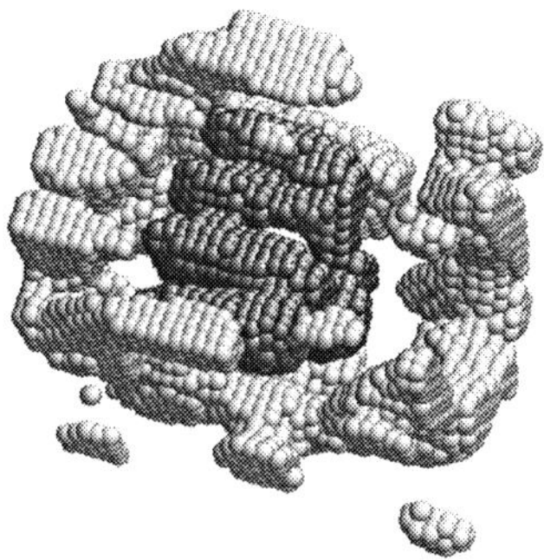

Figure 2. Low-resolution structure of LDL obtained by *ab initio* phasing. The region of high-density values is shown in light gray and the low-density region is in dark gray. The high-density region corresponding to apoB proteins and phospholipids heads form the outer "shell" of the particle. The inner low-density region is formed by lipid layers [16]. (Small spheres forming the image are used as a tool to represent the shape of the particle and have no structural meaning).

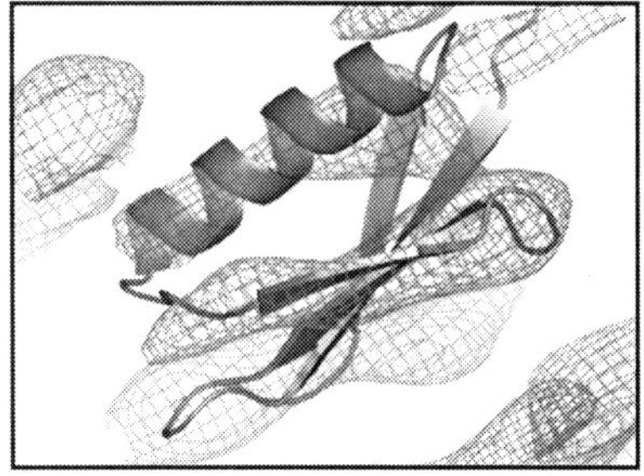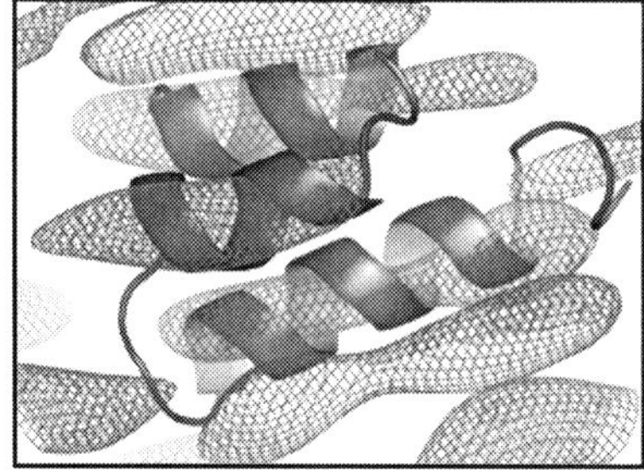

Figure 3. Ab initio phased Fourier syntheses. *Left*: Protein G [5], 8 Å, 85 reflections. *Right*: ER-1 [1], 7.5 Å, 39 reflections.

- An opinion that a few dozens of reflections, in comparison with the thousands of reflections used, do not add any new significant information

To a large extent these reasons are justified and the most of known protein structures were successfully solved in the absence of low-resolution reflections. Nevertheless, there exist cases when the low-resolution data become important:

- Large macromolecules and macromolecular complexes where a low-resolution image has valuable information in itself (e.g., the ribosome, lipoproteins, and large membrane proteins)

- Crystals that do not diffract to high resolution

- Crystals with a mixture of ordered and disordered regions (e.g., protein and nucleic acid components in a virus, lipoproteins, and membrane proteins)

- Cases where the knowledge of the low-resolution envelope is useful to solve the high-resolution phase problem; this includes averaging and density modification methods

- Cases where the low-resolution cutoff causes large image distortions.

It is worthy of noting that nowadays low-resolution *ab initio* phasing methods cannot yet be used routinely, but require a lot of human work, different for different particular objects.

2. Selection criteria

To compensate the lack of experimental information some additional "external" information might be involved in the phasing procedure. In the approach discussed, such information is involved in the form of selection criteria (score functions, figures of merit). The selection criteria reflect expected properties of the true phase set. In "classical" direct methods, such criteria are based mostly on the "atomicity" or "separated atoms" properties of the object studied. They are very effective at near to atomic resolution, but meet difficulties when applied in low- and middle-resolution studies. Several examples of selection criteria developed for low-resolution studies are discussed briefly below, while this list is obviously not exhaustive.

2.1. FEW ATOMS MODEL METHOD

A simple property of a macromolecular crystal is that at low resolution its content may be approximated by the sum of contribution from a few huge dummy atoms ("Gaussian spheres", "globs") so that the structure factors magnitudes calculated from this few atoms model (FAM) are close to the observed ones and phases are close to the true phases [7, 11]. If the atomic coordinates are chosen randomly in the unit cell one can believe that the high correlation of the calculated (from this random FAM) structure factors magnitudes with the observed ones means a high quality of corresponding FAM-calculated phases. Therefore, the value of the magnitude correlation coefficient may serve as a selection criterion to judge expected quality of the corresponding random phase set: a phase set calculated from the given random FAM is rejected if correlation of the magnitudes is low enough [11, 12].

2.2. FOURIER SYNTHESES HISTOGRAMS

This criterion represents another group of selection criteria, where the analyzed object is a Fourier map calculated with the observed magnitudes and trial random phases. The histogram $\{v_k\}, k = 1, \ldots, K$ of a Fourier synthesis $\rho(\mathbf{r})$ indicates which values are present and how frequently these values appear in the synthesis. Let the interval (ρ_{min}, ρ_{max}) of possible values of $\rho(\mathbf{r})$ be divided into K equal parts (bins) and for every bin the frequency $v_k = n_k/N, k = 1, \ldots, K$, be calculated, where N is the total number of grid points at which $\rho(\mathbf{r})$ is calculated and n_k is the number of grid points with $\rho(\mathbf{r})$ values belonging to the kth bin. The set of frequencies $\{v_k\}, k = 1, \ldots, K$ is called the histogram corresponding to the function $\rho(\mathbf{r})$. It was shown that this histogram is sensitive to phase errors. The standard histogram $\{v_k^{exact}\}$, $k = 1, \ldots, K$ is the one corresponding to the Fourier synthesis calculated with the observed magnitudes and the correct phases. The standard histogram at a particular resolution can be predicted before phases are determined [8].

For any randomly generated phase variant the corresponding Fourier synthesis and its histogram can be calculated. The correlation of the calculated and standard histograms (or some other measure of their closeness as well) may be used as an estimate of the quality of the phase set.

2.3. CONNECTIVITY-BASED CRITERION

The regions $\Omega_\kappa = \{\mathbf{r} : \rho(\mathbf{r}) > \kappa\}$ of relatively high values in the properly phased Fourier synthesis consist in a small number of connected components [2]. One can expect to have the number of connected components equal to the number of molecules in the unit cell for low-resolution syntheses and to have a small number of such components in medium-resolution syntheses. For every random phase set the corresponding Fourier synthesis can be calculated and the number of connected regions can be defined. The phase set may be rejected if this number does not correspond to the expectations [13].

2.4. LIKELIHOOD-BASED CRITERION

The Fourier synthesis calculated with the observed magnitudes and trial random phases may be used do define a trial molecular mask $\Omega_\kappa = \{\mathbf{r}:\rho(\mathbf{r}) > \kappa\}$. To estimate the quality of this mask (and by this estimate the quality of the phase set used) the statistical likelihood may be used. The correct molecular region Ω^{exact} contains almost all atoms of the object studied, and at low resolution, models even with random atomic positions inside Ω^{exact} may give a good approximation to the observed magnitudes. On the other hand, if the region Ω is chosen arbitrarily, then the possibility of reproducing the observed magnitudes when placing atoms randomly inside such region is low. Therefore, the likelihood, i.e., the probability to reproduce the observed magnitudes by values $\{F^{\text{calc}}(\mathbf{s})\}$ calculated from the atoms randomly placed in a trial molecular region, may serve as a selection criterion [3, 12]. The phase variant that leads to the mask possessing of low likelihood must be rejected [20].

3. Multifiltering phasing procedure

The criteria discussed above (as well as some other criteria tried) reveal general weakness when applied at low resolution [17]:

- Best value of a selection criterion may correspond to a quite bad phase set
- Value of the selections criterion for a good phase set may be significantly worse than the best selection criterion value
- Local refinement may lead to a very good selection criterion value without any improvement of the phases

These properties make an attempt to find the solution of the low-resolution phase problem by means of minimization (or maximization) of some selection criterion highly unreliable. At the same time the collection of phase sets selected on the base of a proper criterion from an ensemble of randomly generated phase sets has a higher concentration of good variants in comparison with the random ensemble. We call this property as enrichment and use it as a base for multifiltering phasing procedure.

In this paper, we discuss an approach to low-resolution phasing based on the multifiltering procedure [10, 14, 17]. At every cycle of this procedure:

- Large number of random phase sets are generated
- Generated ensemble of phase sets is filtered applying different selection rules
- Phase sets that fit all selection rules (the selected phase sets) form the output of the cycle

The output is used to produce estimates of phase values and to modify the mode of generation of random phases at the next cycle. A key feature of this procedure is that at every its cycle we do not try to find one or a few phase sets that fit the best possible the selection rules but on the contrary we remove definitely bad phase sets. In fact, this multifiltering phasing approach may be useful with necessary modifications in different situations and not only at low resolution.

The cycle is repeated several times varying the random phase-generating mode, selection rules, the set of reflections to be phased, etc.

4.   Generation of an ensemble of random phase sets

The generation of a large number of random phase sets constitutes the first step of the phasing cycle and defines how the configuration space of all phase sets is explored. Two different ways to generate random variants are used. In the first one, all phases are considered directly as independent random variables. In the absence of any phase information all phase values are considered as equally probable. When some phase information for a reflection **s** is available, it may be used to generate random phase values distributed accordingly to a given probability distribution, e.g., unimodal von Mises ("circular normal", "Sim", etc.) distribution

$$P(\varphi) = \frac{1}{I_0(t)}\exp\left[t\cos(\varphi - \theta)\right] = \Omega(A, B)\exp\left[A\cos\varphi + B\sin\varphi\right],$$

or bimodal Hendrickson–Lattman (HL) distribution

$$P(\varphi) = \Omega(A, B, C, D)\exp\left[A\cos\varphi + B\sin\varphi + C\cos 2\varphi + D\sin 2\varphi\right].$$

Obviously, the space group symmetry restrictions on the possible phase values must be taken into account when generating random phases.

In another way, the phases are calculated through some model with random parameters. For example, a phase variant may be calculated from a model composed from a small number of huge "globs" whose coordinates are considered now as primary random values [11]. The coordinates may be distributed uniformly in the whole unit cell at the very beginning of phasing and inside some molecular mask later when some phase information becomes available. A more advanced example is calculation of phase values with the use of atomic model of homologous object with randomly chosen rotation and translation parameters.

5.   Processing of the selected phase sets

Some simplest steps of processing of the output of the cycle of phasing are discussed below. More sophisticated approaches may be used as well.

5.1. ALIGNMENT OF PHASE SETS

Two phase sets, apparently different, may result in Fourier syntheses that differ only by an origin shift permitted for the given space group (and/or enantiomorph choice). For example, for any vector $\mathbf{t}$, if two syntheses are calculated using the same magnitudes $\{F^{obs}(\mathbf{s})\}, \mathbf{s} \in S$, but different phase sets $\{\varphi_1(\mathbf{s})\}$ and $\{\varphi_2(\mathbf{s})\}$ with $\varphi_2(\mathbf{s}) = \varphi_1(\mathbf{s}) + 2\pi(\mathbf{s},\mathbf{t})$, then the only difference in the syntheses is the shift of one of them by vector $\mathbf{t}$ with respect to another. Such two phase sets must be considered as equivalent ones and any selection criterion should give them the same score values. At the same time, results of direct comparison or averaging of these two sets are unpredictable. To calculate a proper estimate of the closeness of two phase variants the map *alignment* with respect to the choice of the origin and enantiomer must be performed [9], i.e., the translation vector $\mathbf{t}^*$ and the sign $\theta = \pm 1$ must be found that make the differences $\varphi_1(\mathbf{s}) - [\theta \cdot \varphi_2(\mathbf{s}) - 2\pi(\mathbf{s},\mathbf{t})]$ for $\mathbf{s} \in S$ as small as possible, and the comparison and averaging must be applied to the phase sets $\{\varphi_1(\mathbf{s})\}$ and $\{\varphi_2{}^*(\mathbf{s})\}$ where $\varphi_2{}^*(\mathbf{s}) = \theta \cdot \varphi_2(\mathbf{s}) - 2\pi(\mathbf{s}, \mathbf{t}^*)$. This procedure of the phase alignment is especially important at first stages of phasing when any phase values may be generated.

If the number of reflections is very small, the problem of alignment may be avoided to some extent by fixing the phase values for several origin and enatiomorph setting reflections.

Two figures are mainly used to estimate the similarity of two phase sets $\mathbf{v}_1 = \{\varphi_1(\mathbf{s})\}$ and $\mathbf{v}_2 = \{\varphi_2(\mathbf{s})\}$. The mean phase difference appeals to the phase values only

$$D_\varphi = \frac{1}{M} \sum_{s \in S} \left| \varphi_1(\mathbf{s}) - \varphi_2(\mathbf{s}) \right|$$

where M is the number of reflection in the set S. This figure is strongly influenced by a number of weak reflections with badly defined phases.

The map correlation coefficient is defined as

$$C_\varphi(\mathbf{v}_1, \mathbf{v}_2) = \frac{\int \left(\rho_1(\mathbf{r}) - \langle\rho_1\rangle\right)\left(\rho_2(\mathbf{r}) - \langle\rho_2\rangle\right) dV_r}{\sqrt{\int \left(\rho_1(\mathbf{r}) - \langle\rho_1\rangle\right)^2 dV_r \int \left(\rho_2(\mathbf{r}) - \langle\rho_2\rangle\right)^2 dV_r}}$$

$$= \sum_{s \in S} F^{obs}(\mathbf{s})^2 \cos\left(\varphi_1(\mathbf{s}) - \varphi_2(\mathbf{s})\right) \Big/ \sum_{s \in S} F^{obs}(\mathbf{s})^2$$

where $\rho_1(\mathbf{r})$ and $\rho_2(\mathbf{r})$ are Fourier syntheses calculated with the observed magnitudes and phases $\mathbf{v}_1 = \{\varphi_1(\mathbf{s})\}$ and $\mathbf{v}_2 = \{\varphi_2(\mathbf{s})\}$ correspondingly, and the sums in reciprocal space are calculated without $F(0)$ term. The map correlation coefficient is influenced mostly by phases of the strongest low-resolution reflections and is not sensitive to the phases of weak and intermediate reflections. The calculation of the map correlation coefficient in resolution shells compensates to some extent this shortcoming.

5.2. AVERAGING

The simplest processing of a selected phase sets is their averaging. For every particular reflection **s** from the set S the *best* phase and its *figure of merit* are calculated as

$$m(\mathbf{s})\exp\left[i\varphi^{\text{best}}(\mathbf{s})\right] = \frac{1}{K}\sum_{j=1}^{K}\exp\left[i\varphi_j(\mathbf{s})\right],$$

where $\varphi_j(\mathbf{s})$ is the phase in jth selected variant corresponding to the reflection **s** and K is the number of selected variants. The figure of merit reflects the divergence of the phases corresponding to the same reflection in different selected sets. It is worth noting that

$$m(\mathbf{s}) = \frac{1}{K}\sum_{j=1}^{K}\cos\left(\varphi_j(\mathbf{s}) - \varphi^{\text{best}}(\mathbf{s})\right)$$

so that $m(\mathbf{s}) = 1$ if the phase for this reflection is the same in all selected phase variants and $m(\mathbf{s}) \approx 0$ if the phases are distributed almost uniformly in $[0, 2\pi]$ interval. The found m, φ^{best} values may be used to define von Mises probability distribution for the next cycle of random phase generation supposing $\theta = \varphi^{\text{best}}$ and t parameter value defined from the equation $I_1(t)/I_0(t) = m$.

5.3. ASSIGNING THE HENDRICKSON–LATTMAN PROBABILITY DISTRIBUTIONS

The output of the phasing cycle may be used to derive for every reflection an approximate *HL* distribution of the phase. Let $\{\varphi_j\}$, $j = 1, \ldots, M$ are the phases corresponding to the same reflection **s** in different selected phase sets. Then the coefficients of the corresponding to this reflection HL distribution may be defined, e.g., by maximization of the likelihood $L(A,B,C,D)$ or by minimization of

$$-\ln L\left(A,B,C,D\right) = -\sum_{j=1}^{M}\ln P\left(\varphi_j\right)$$

$$= -M\ln\Omega\left(A,B,C,D\right) - A\sum_{j=1}^{M}\cos\varphi_j - B\sum_{j=1}^{M}\sin\varphi_j$$

$$- C\sum_{j=1}^{M}\cos 2\varphi_j - D\sum_{j=1}^{M}\sin 2\varphi_j$$

5.4. CLUSTER ANALYSIS

An accurate treatment of a selected population involves methods of cluster analysis [10, 11]. Cluster analysis is a developed branch of applied mathematics aimed to separate a set of points in a multidimensional space into several compact groups of points called "clusters" (or "classes") so that the points inside a particular cluster are close to each other while different clusters are distanced in space. Methods of cluster analysis use the matrix of point-to-point distances as input information. The mean phase difference in two previously *aligned* phase set (as well as other measures for phase sets closeness) may be used to calculate such matrix in our case. If the cluster analysis shows that the selected population can be divided into several significantly different clusters, then the averaging is performed in each cluster separately, resulting in several alternatives for the solution of the phase problem. This creates a sort of branching in phasing requiring multisolution strategies.

Acknowledgments

VYL and NL were supported by RFBR grant 03-04-48155. A part of this work was supported by the CNRS–RAS collaboration. AU thanks Pole "Intelligence Logicielle" and CRVHP, LORIA, and Nancy for financial support that made this publication possible. AU is member of GdR 2417 CNRS. The authors thank E. Vernoslova, T. Petrova, T. Skovoroda, A. Podjarny, M. Baumstark, S. Ritter, K. Diederichs, M. Pos, M. Seeger, and A. Fokine for their contribution to different parts of the work presented in this article. The program *PyMOL* [4] was used to show maps.

References

1. Anderson, D., Weiss, M., and Eisenberg, D. (1996) A challenging case for protein structure determination: the mating pheromone Er-1 from Euplotes raikovi. *Acta Crystallographica*, **D52**: 469–480.
2. Baker, D., Krukowski, A.E., and Agard, D.A. (1993) Uniqueness and the ab initio phase problem in macromolecular crystallography. *Acta Crystallographica*, **D49**: 186–192.
3. Bricogne, G., and Gilmore, C.J. (1990) A multisolution method of phase determination by combined maximization of entropy and likelihood. I. Theory, algorithms and strategy. *Acta Crystallographica*, **A46**: 284–297.
4. DeLano, W.L. (2002) The PyMOL Molecular Graphics System. http://www.pymol.org
5. Derrick, J.P., and Wigley, D.B. (1994) The third IgG-binding domain from Streptococcal protein G. An analysis by X-ray crystallography of the structure alone and in a complex with Fab. *Journal of Molecular Biology*, **243**: 906–918.
6. Gilmore, C.J. (2000) Direct methods and protein crystallography at low resolution. *Acta Crystallographica*, **D56**: 1205–1214.
7. Guo, D.Y., Blessing, R.H., and Langs, D.A. (2000) Globbic approximation in low-resolution direct-methods phasing. *Acta Crystallographica*, **D56**: 1148–1155.

8. Lunin, V.Yu. (1993) Electron-density histograms and the phase problem. *Acta Crystallographica*, **D49**: 90–99.

9. Lunin, V.Y., and Lunina, N. (1996) The map correlation coefficient for optimally superposed maps. *Acta Crystallographica*, **A52**: 365–368.

10. Lunin, V.Y., Urzhumtsev, A., and Skovoroda, T. (1990) Direct low-resolution phasing from electron-density histograms in protein crystallography. *Acta Crystallographica*, **A46**: 540–544.

11. Lunin, V.Y., Lunina, N., Petrova, T., Vernoslova, E., Urzhumtsev, A., and Podjarny, A. (1995) On the *ab initio* solution of the phase problem for macromolecules at very low resolution: the few atoms model method. *Acta Crystallographica*, **D51**: 896–903.

12. Lunin, V.Y., Lunina, N., Petrova, T., Urzhumtsev, A., and Podjarny, A. (1998) On the *ab initio* solution of the phase problem for macromolecules at very low resolution. II. Generalized likelihood based approach to the cluster discrimination. *Acta Crystallographica*, **D54**: 726–734.

13. Lunin, V.Y., Lunina, N., and Urzhumtsev, A. (2000) Topological properties of high density regions and *ab initio* phasing at low resolution. *Acta Crystallographica*, **A56**: 375–382.

14. Lunin, V.Y., Lunina, N., Petrova, T., Skovoroda, T., Urzhumtsev, A., and Podjarny, A. (2000) Low resolution *ab initio* phasing. Problems and advances. *Acta Crystallographica*, **D56**: 1223–1232.

15. Lunin V.Y., Urzhumtsev A., and Podjarny A.D. (2001) Phasing at low resolution: methods and applications. In *Advances in Structure Analysis*. Edited by Kuzel, R., and Hasek, J. Chech and Slovak Crystallographic Association, pp. 4–36.

16. Lunin, V.Y., Lunina, N., Ritter, S., Frey, I., Keul, J., Diederichs, K., Podjarny, A., Urzhumtsev, A., and Baumstark, M. (2001) Low-resolution data analysis for the low-density lipoprotein particle. *Acta Crystallographica*, **D57**: 108–121.

17. Lunin, V.Y., Lunina, N., Podjarny, A., Bockmayr, A., and Urzhumtsev, A. (2002) Ab initio phasing starting from low resolution. *Zeitschrift für Kristallographie*, **217**: 668–685.

18. Lunina, N.L., Lunin, V.Y., Pos, K.M., Seeger, M.A., Diederichs, K., and Baumstark M.W. (2005) "On site" direct phasing of a membrane protein "AcrB" at low resolution. Submitted to ESRF Highlights.

19. Murakami, S., Nakashima, R., Yamashita, E., and Yamagushi, A. (2002) Crystal structure of bacterial multidrug efflux transporter AcrB. *Nature*, **419**: 587–593.

20. Petrova, T., Lunin, V.Y., and Podjarny, A. (2000) Ab initio low-resolution phasing in crystallography of macromolecules by maximization of likelihood. *Acta Crystallographica*, **D56**: 1245–1252.

21. Sheldrick, G.M., Hauptman, H.A., Weeks, C.M., Miller, R., and Usón, I. (2001) Direct methods. In *International Tables for Crystallography*, vol. F. Edited by Rossmann, M., and Arnold, E. Dordrecht/Boston/London: Kluwer Academic, pp. 333–345.

22. Urzhumtsev, A.G., Lunina, N.L., Skovoroda, T.P., Podjarny, A.D., and Lunin, V.Y. (2000) Density constraints and low resolution phasing. *Acta Crystallographica*, **D56**: 1233–1244.

STRUCTURAL GENOMICS OF *MYCOBACTERIUM TUBERCULOSIS*: A SEARCH FOR FUNCTION AND NEW DRUG TARGETS

TED BAKER

Center for Molecular Biodiscovery and School of Biological Sciences, University of Auckland, Auckland, New Zealand and International TB Structural Genomics Consortium

Abstract: Structural genomics initiatives have various goals: from the discovery of new folds to providing representative structures for all protein families, to the discovery of function from structure, and the characterization of new drug targets. The TB Structural Genomics Consortium (TBSGC), an international collaboration of more than 50 laboratories, was formed to address the worldwide problem of TB, through its focus on *Mycobacterium tuberculosis*, the causative agent. The goals are to characterize new drug targets and to gain a deeper understanding of TB biology. The project has now entered a very productive phase, with more than one third of the genes cloned, many proteins purified, and more than 100 structures determined. Some of these are for already-validated drug targets. Others have led to new functional understanding of important processes in the biology of the organism, providing validation of the structure-to-function paradigm.

Keywords: structural genomics; *Mycobacterium tuberculosis*; tuberculosis; drug targets; TB biology.

1. Introduction

The increasing availability of complete genome sequences from a variety of living systems has begun a revolutionary change in the biological sciences. New opportunities now exist for addressing old problems, some questions can be asked that could not even be formulated before, and many new "high-throughput" technologies are being developed to permit genome-scale approaches.

One of the most striking outcomes of genome sequence analysis has been the realization that our current understanding of the functions of genes and their protein products is far from complete. The figures for the genome

R.J. Read and J.L. Sussman (eds.): Evolving Methods for Macromolecular Crystallography, 135–144
© 2007. *Springer*

sequence of the bacterium that causes TB, *Mycobacterium tuberculosis* (*Mtb*), are typical of most genome sequences. In the initial functional annotation, functions were attributed to ~40% of gene products, some information or similarity could be found for another 44%, many of which were "conserved hypotheticals" (meaning that they were found also in other organisms, but were of unknown function) and 16% were completely unknown, being found only in *Mtb* or in other mycobacteria [1]. Most functions are inferred by homology, meaning that a homologous protein in another organism has been functionally characterized but the *Mtb* protein has not. Others are "low-resolution" functions, where, for example, a protein may be described as a short-chain dehydrogenase or transcription factor, without knowing what its specific substrate or role is. A conservative estimate is that at least 65% of gene products are of unknown or uncertain function. Lastly, there are clearly many unknown biochemical pathways or pathways that differ in detail from those in other organisms.

A variety of experimental and bioinformatic approaches are now being developed to meet the challenges of interpreting and exploiting the knowledge that is hidden or poorly recognized in genome sequences. These include the use of genome-wide microarrays to address gene transcription and expression [2, 3], and the launching of a number of structural genomics initiatives, the subject of this presentation.

2.   Structural genomics concepts and goals

Structural genomics, or perhaps more correctly structural proteomics, is based on two overarching concepts. Firstly, protein classification schemes such as SCOP (http://scop.mrc-lmb.cam.ac.uk/scop/) and CATH (http://www.biochem. ucl.ac.uk/bsm/cath) make clear that proteins have only a limited number of different folds, perhaps little more that 1,000, depending on what is meant by "different". They may also be grouped into superfamilies, which are assumed to have arisen by evolutionary divergence, and within which are recognizable sequence families that share significant levels of sequence identity. If representative structures were available for all sequence families, other family members could be modeled by homology modeling. Secondly, since the function of a protein depends on its three-dimensional (3D) structure, it may be possible to discover the function of a protein, or derive important clues, from knowledge of the 3D structure. An added benefit is that the structure would be available to direct future biochemical assays or to use for drug development.

The term structural genomics is thus given to initiatives that use complete genome sequences as the starting point for large-scale protein structure determination [4–7]. The goals of such initiatives are varied, and include:

- Development of technologies and methods for large-scale or more rapid or efficient protein structure determination
- Discovery of new folds, to establish the extent of "fold space" [8]
- Determination of structures for the complete proteome of a model organism
- Determination of representative 3D structures for all sequence families, to place other family members within modeling distance [9]
- Discovery of function from structure [10, 11]
- Characterization of potential new drug targets identified from genome analysis

A number of structural genomics enterprises have been initiated in the past 4–5 years. The largest of these are the National Institutes of Health (NIH)-supported Protein Structure Initiative in the USA, the Japanese Protein 3000 project, and the European SPinE project. Many smaller projects, with more targeted goals, have been initiated in other countries, and an International Structural Genomics Organization has been formed (http://www.isgo.org/) to promote dialogue between these efforts.

3. Technologies

Structural genomics poses much greater challenges than DNA sequencing for high-throughput operations, due to the complexity of the steps involved. Each target must move through a pipeline involving: gene cloning (by polymerase chain reaction [PCR]); expression of soluble proteins in a suitable host; purification, crystallization; and structure determination by X-ray crystallography or nuclear magnetic resonance (NMR). In the larger structural genomics initiatives, these steps have been automated through the use of robotics, for the protein production and crystallization steps [12], and program suites that automate many aspects of structure determination [13].

Many of the robotic technologies are currently not feasible for smaller laboratories, for reasons of cost or scale. However, the development of improved expression vectors and protocols [14], including the wide adoption of recombination-based cloning, new approaches to solubilization and crystallization, automation of synchrotron beam lines, and more powerful software, benefit the whole community. In our own laboratory, we have adopted the Gateway cloning system, with multipipettors and 96-well plates, for rapid expression testing. Use of a nanoliter crystallization robot has markedly improved success in protein crystallization [15]; its speed and accuracy, and the use of smaller (100 nL) volumes mean faster drop equilibration, less chance of protein degradation, and the ability to use only the best protein fraction for crystallization. Powerful automated methods for phasing [16, 17], map improvement [18], and model-building [18, 19] developed in the past few years mean that it is not uncommon for a structure to be solved, and model largely built, within 24 h of MAD or SAD data collection at a synchrotron.

4. The TB Structural Genomics Consortium

The TB Structural Genomics Consortium (TBSGC) [20, 21] was formed in 2000 as one of the seven original initiatives funded by the US National Institutes of Health under their Protein Structure Initiative. However, the TBSGC (http://www.tbgenomics.org/) differs from the other PSI consortia in both its goals and its mode of operation. The focus is on *M. tuberculosis* (*Mtb*), the cause of TB. Due to the global impact of this organism, which causes more deaths than any other infectious agent (~2 million people annually) and is estimated to infect one third of the world's population, the TBSGC has chosen to operate in a globally inclusive manner. Only the "core" US groups are funded by NIH. Other groups from around the world can participate, however, with their own funding, sharing the core US facilities and attending annual retreats. All consortium members accept agreed rules of operation including coordination of targeting to reduce duplication of effort, openness in the recording of experimental progress in a shared database, sharing of results, and prompt deposition of results in public databases. At the present time more than 50 centers in nine countries participate in this unique international effort.

5. Targets and goals

Given the importance of TB as a world health threat, the focus of the TBSGC is firmly on function [20, 21]. Although there are several effective anti-TB drugs (isoniazid, ethambutol, pyrazinamide, and rifampicin), treatment of *Mtb* infection is made difficult by the curious nature of this highly specialized human pathogen [22]. First, its thick and waxy cell wall is rich in unusual lipids, glycolipids, and polysaccharides, some of which contribute to mycobacterial longevity, trigger host reactions, and act in pathogenesis [1]. Second, it has the unusual ability to enter a dormant or persistent phase after engulfment by host macrophages, remaining in this state for many years, but able to reemerge as an active infection when immunity wanes [22]. Current treatments are long (6–9 months, with multiple drugs) and resistance is emerging.

Targets for 3D structure determination are chosen with two main goals in mind: the characterization of potential new drug targets, and the investigation of aspects of TB biology. Many of the drug targets are classic biosynthetic enzymes that are easily recognized in the genome. Others are suggested by biological studies. For example, a genome-wide transposon mutagenesis study [23] has identified a large cohort of genes that are essential for growth of *Mtb*; many of these correspond with classic biosynthetic targets, but others are of completely unknown function. Likewise, microarrray studies that have examined the upregulation or downregulation of gene expression in response to antibiotic challenge [2] or to conditions of hypoxia (thought to replicate the onset of dormancy) [3] have implicated further gene products of unknown function. These proteins are prime candidates for structural analysis to address function because of their evident (but unexplained) importance in *Mtb* biology. Some of them will be the drug targets of the future.

6. Progress to date

Output from the TBSGC mirrors the step-by-step attrition that is typical of structural genomics enterprises, with less than 100% success at each step resulting in a final yield of structures that is less than 10% overall. The current figures are 1,269 genes cloned (about one third of the genome), 982 expressed, 426 soluble, 368 purified, 156 crystallized, and 93 structures. However, the figures themselves do not do justice to the extent of the achievement. Targets were chosen for their functional importance rather than their tractability. The solubility of *Mtb* proteins expressed in *Escherichia coli* proved to be major challenge, and one that is still not overcome. Most of the protein structures have been determined in the last 2 years, once the establishment phase was complete, and many more will come in the next two.

Importantly, many of the protein structures determined within the TBSGC are valid drug targets [24]. Some are from classic biosynthetic pathways that are common to many bacteria, but are not present in humans. Examples include LeuA (α-isopropylmalate synthase) and LeuB (3-isopropylmalate dehydrogenase), both from the leucine biosynthesis pathway [25, 26]; LysA (diaminopimelate decarboxylase) [27] from the lysine biosynthetic pathway; and HisG (ATP phosphoribosyltransferase) [28] from the pathway for histidine biosynthesis. All of these have been shown from gene knockouts to be essential for *Mtb* growth. Cofactor biosynthesis is another fertile area, and here, for example, structures, have been determined for the enzymes that catalyze the first and last steps in the biosynthesis of pantothenate (vitamin B5), an essential precursor for the biosynthesis of coenzyme A: PanB (ketopantoate hydroxymethyltransferase) [29] and PanC (pantothenate synthase) [30].

Other targets are more specific to *Mtb*. The enzymes malate synthase and isocitrate lyase form the so-called glyoxylate shunt that is critical for survival in the persistent phase of *Mtb* infection. Structures have been determined for both these enzymes [22], as the basis for ongoing drug development. Structures have been determined for several mycolic acid cyclopropane synthase enzymes [22] that catalyze key modifications to the cell wall mycolic acids; these are important for both pathogenesis and persistence. Secreted proteins form another group of important targets due to their likely role in host–pathogen interactions. These also have the major advantage, for drug development, of being extracellular. Examples include the mycolyl transferases, also called antigens 85A, 85B, and 85C [31, 32], and an adaptin-like secreted protein of unknown function MPT63 [33].

Finally, a number of intriguing observations from analysis of the *Mtb* genome [1] are only now beginning to be understood. Structural information is emerging [34] for a group of 11 putative serine/threonine kinases that could play a key role in host–pathogen interactions. More recalcitrant is the large complement of gene products referred to as PE and PPE proteins (for their Pro-Glu and Pro-Pro-Glu repeats), none of which has yet been crystallized.

7. Function

Where proteins are of unknown or uncertain function, structural analyses can provide a variety of outcomes, ranging from very little new information (other than the structure) to a strong functional prediction that can be tested by experiment [10, 11, 35]. Functional clues can come from a variety of sources:

- The fold of the protein may place it within a known family or superfamily despite the lack of any clear indication from its amino acid sequence

- Recognition of a catalytic motif may suggest a particular activity

- A bound cofactor, metabolite, or solvent molecule may point to a functionally important site

- Sequence conservation may indicate a likely active site

A few examples from our own laboratory will serve to illustrate some of these possibilities.

- RV1170. This protein was originally annotated as "conserved hypothetical." After the publication of the *Mtb* genome sequence, however, experimental studies in the related organism *Mycobacterium smegmatis* identified its homolog in that organism as being involved in the biosynthesis of mycothiol, a small-molecule reductant that plays an equivalent role to glutathione. From the structure of Rv1170 [36, 37], the active site was identified by the presence of a conserved AHPDDE motif, adjacent to a bound metal ion. Importantly, an adventitiously bound molecule of the detergent β-octylglucoside provided a nice model for a bound substrate since the glucosyl group modeled the *N*-acetylglucosyl moiety of mycothiol.

- Rv1347c. This protein was annotated as an aminoglycoside *N*-acetyltransferase (AAC), with a supposed activity of acetylating amino groups on aminoglycoside antibiotics such as streptomycin and kanamycin, thereby inactivating them. Its sequence identity with known enzymes of this type was only 15%, however, and no such activity could be demonstrated. The crystal structure [38] showed that the fold clearly identified Rv1347c as belonging to the GCN5-related *N*-acetyltransferase (GNAT) family, to which the AACs belong. Further investigation showed, however, that it is an essential gene in *Mtb*; its expression is regulated by iron; its closest homologs are other bacterial proteins involved in the biosynthesis of siderophores (small molecule chelators used to acquire iron); and its neighbors in the *Mtb* genome are implicated in the biosynthesis of mycobactin, used by *Mtb* to take up iron. Crucially, in the crystal structure "extra" density attributed to detergent marked a hydrophobic channel leading to the active site. The conclusion, which has since been verified experimentally [39], was that Rv1347c is a "missing" enzyme of mycobactin biosynthesis that adds a long-chain acyl group to the *N*-hydroxylysine side chain of mycobactin.

- Rv3853. This protein was originally annotated as MenG, an *S*-adenosylmethionine (SAM)-dependent methyltransferase in the biosynthesis of menaquinone. The gene was remote from other menaquinone biosynthesis genes in the genome, however, and no methyltransferase activity could be demonstrated. The crystal structure analysis showed that the fold of Rv3853 is completely different from that of any known methyltransferase, and the conclusion is that this gene has been annotated wrongly [40]. This incorrect annotation is propagated through many bacterial genomes in which homologs exist. Unfortunately, the fold of MenG does not place it clearly in any known family, and although several solvent molecules are bound in sites that could be functionally important, they do not indicate a testable function.

- PAE2754. This protein is from a hyperthermophilic archaeon, *Pyrobaculum aerophilum*, but was chosen because it was homologous with four *Mtb* genes (Rv0065, Rv0549, Rv0960, and Rv1720), all of unknown function, classified in Pfam as PIN domains. The crystal structure revealed a tetramer in which each monomer has four acidic amino acids clustered in a pocket. A search of other genome sequences finds many similar proteins, in which the four acidic amino acids are strictly conserved, suggesting that they are at the active site, and involved in metal binding. A weak similarity was found between the fold of PAE2754, and that of the T4 RNase H family of Mg^{2+}-dependent nucleases. Subsequent biochemical assays showed that PAE2754 is indeed a nuclease, a function that is likely to be shared by its many homologs [41]. On delving further, however, it was discovered that *Mtb* has no fewer than 48 homologs of PAE2754, and that 38 of these are found paired in the genome with antitoxin-like genes [42]. This leads to the intriguing hypothesis that these are toxin–antitoxin pairs, and that their extraordinary expansion in *Mtb* points to an important biological role in growth or survival, perhaps in dormancy.

8. Perspectives

In 2000, at the time of the founding of the TBSGC only eight protein structures from *M. tuberculosis* were available. Today, more than 200 can be found in the Protein Data Bank and although some are duplicates or ligand complexes, structures are now available for more than 100 unique *Mtb* proteins. About two thirds of these were determined within the TBSGC, and it is clear that knowledge of the *Mtb* genome sequence has stimulated many research groups worldwide, inside or outside the TBSGC. Many of the structures are for proteins that are valid drug targets, shown by gene knockout to be essential for *Mtb* growth. The availability of these structures, their cloned genes, and protein production protocols, will be of huge benefit for combating this devastating pathogen. Some practical challenges remain. Up to now, typically only 30–40% of *Mtb* proteins are obtained in soluble form when expressed in *E. coli*, even after the exclusion of membrane proteins. Some may be rescued by using alternative hosts, or coexpressing with interacting partners. For the full impacts on *Mtb* biology to be felt, however, a much greater focus on protein–protein complexes and membrane proteins is needed.

Acknowledgments

The work described here was supported by the US National Institutes of Health, through support of the TB Structural Genomics Consortium under the Protein Structure Initiative, and by the Health Research Council of New Zealand, the New Economy Research Fund of New Zealand, and the Centre for Molecular Biodiscovery, University of Auckland.

References

1. Cole, S.T. et al. (1998) *Nature*, **393**: 537–544.
2. Wilson, M., DeRisi, J., Kristensen, H.-H., Imboden, P., Rane, S., Brown, P.O., and Schoolnik, G.K. (1999) *Proceedings of the National Academy of Sciences of the USA*, **96**: 12833–12838.
3. Sherman, D.R., Voskuil, M., Schnappinger, D., Liao, R., Harrell, M.I., and Schoolnik, G.K. (2001) *Proceedings of the National Academy of Sciences of the USA*, **98**: 7534–7539.
4. Terwilliger, T.C., Waldo, G., Peat, T.S., Newman, J.M., Chu, K., and Berendzen, J. (1998) *Protein Science*, **7**: 1851–1856.
5. Burley, S.K. et al. (1999) *Nature Genetics*, **23**: 151–157.
6. (2000) Various articles in *Nature Structural Biology*, **7**(suppl.).
7. Mittl, P.R.E. and Grutter, M.G. (2001) *Current Opinion in Chemical Biology*, **5**: 402–408.
8. Brenner, S.E. and Levitt, M. (2000) *Protein Science*, **9**: 197–200.
9. Vitkup, D., Melamud, E., Moult, J., and Sander, C. (2001) *Nature Structural Biology*, **8**: 559–566.
10. Teichmann, S.A., Murzin, A.G., and Chothia, C. (2001) *Current Opinion in Structural Biology*, **11**: 354–363.
11. Zhang, C. and Kim, S.-H. (2003) *Current Opinion in Chemical Biology*, **7**: 1–5.
12. Lesley, S.A. et al. (2002) *Proceedings of the National Academy of Sciences of the USA*, **99**: 11664–11669.
13. Holton, J. and Alber, T. (2004) *Proceedings of the National Academy of Sciences of the USA*, **101**: 1537–1542.
14. Stevens, R.C. (2000) *Structure*, **8**: R177–R185.
15. Sulzenbacher et al. (2002) *Acta Crystallographica*, **D58**: 2109–2115.
16. Terwilliger, T.C. and Berendzen, J. (1999) *Acta Crystallographica*, **D55**: 849–861.
17. Uson, I. and Sheldrick, G.M. (1999) *Current Opinion in Structural Biology*, **9**: 643–648.
18. Terwilliger, T.C. (2000) *Acta Crystallographica*, **D56**: 965–972.
19. Perrakis, A., Morris, R., and Lamzin, V. (1999) *Nature Structural Biology*, **6**: 458–463.
20. Goulding, C.W. et al. (2002) *Current Drug Targets – Infectious Disorders*, **2**: 121–141.
21. Terwilliger, T.C. et al. (2003) *Tuberculosis*, **83**: 223–249.
22. Smith, C.V., Sharma, V., and Sacchettini, J.C. (2004) *Tuberculosis*, **84**: 45–55.
23. Sassetti, C.M., Boyd, D.H., and Rubin, E.J. (2003) *Molecular Microbiology*, **48**: 77–84.
24. Smith, C.V. and Sacchettini, J.C. (2003) *Current Opinion in Structural Biology*, **13**: 658–664.
25. Koon, N., Squire, C.J., and Baker, E.N. (2004) *Proceedings of the National Academy of Sciences of the USA*, **101**: 8295–8230.
26. Singh, R.K., Kefala, G., Janowski, R., Mueller-Dieckmann C., von Kries, J.P., and Weiss, M.S. (2005) *Journal of Molecular Biology*, **346**: 1–11.
27. Gokulan, K., Rupp, B., Pavelka, M.S., Jacobs, W.R., and Sacchettini, J.C. (2003) *Journal of Biological Chemistry*, **278**: 18588–18596.
28. Cho, Y., Sharma, V., and Sacchettini, J.C. (2003) *Journal of Biological Chemistry*, **278**: 8333–8339.
29. Chaudhuri, B.N., Sawaya, M.R., Kim, C.-Y., Waldo, G.S., Park, M.S., Terwilliger, T.C., and Yeates, T.O. (2003) *Structure*, **11**: 753–764.
30. Wang, S. and Eisenberg, D. (2003) *Protein Science*, **12**: 1097–1108.
31. Ronning, D.R., Klabunde, T., Besra, G.S., Vissa, V.D., Belisle, J.T., and Sacchettini, J.C. (2000) *Nature Structural Biology*, **7**: 141–146.
32. Anderson, D.H., Harth, G., Horwitz, M.A., and Eisenberg, D. (2001) *Journal of Molecular Biology*, **307**: 671–681.
33. Goulding, C.W., Parseghian, A., Sawaya, M.R., Cascio, D., Apostol, M.I., Gennaro, M.L., and Eisenberg, D. (2002) *Protein Science*, **11**: 2887–2893.
34. Young, T.A., Delagoutte, B., Endrizzi, J.A., Falick, A.M., and Alber, T. (2003) *Nature Structural Biology*, **10**: 168–174.

35. Eisenstein, E., Gilliland, G., Herzberg, O., Moult, J., Orban, J., Poljak, R.J., Banerjei, L., Richardson, D., and Howard, A.J. (2000) *Current Opinion in Biotechnology*, **11**: 25–30.
36. McCarthy, A.A., Peterson, N.A., Knijff, R., and Baker, E.N. (2004) *Journal of Molecular Biology*, **335**: 1131–1141.
37. Maynes, J.T., Garen, C., Cherney, M.M., Newton, G., Arad, D., Av-Gay, Y., Fahey, R.C., and James, M.N.G. (2003) *Journal of Biological Chemistry*, **278**: 47166–47170.
38. Card, G.L., Peterson, N.A., Smith, C.A., Rupp, B., Schick, B.M., and Baker, E.N. (2005) *Journal of Biological Chemistry*, **280**: 13978–13986.
39. Krithika, R., Marathe, U., Saxena, P., Ansari, M.Z., Mohanty, D., and Gokhale, R.S. (2006) *Proceedings of the National Academy of Sciences of the USA*, **103**: 2069–2074.
40. Johnston, J.M., Arcus, V.L., Morton, C.J., Parker, M.W., and Baker, E.N. (2003) *The Journal of Bacteriology*, **185**: 4057–4065.
41. Arcus, V.L., Backbro, K., Roos, A., Daniel, E.L., and Baker, E.N. (2004) *Journal of Biological Chemistry*, **279**: 16471–16478.
42. Arcus, V.L., Rainey, P.B., and Turner, S.J. (2005) *Trends in Microbiology*, **13**: 360–365.

THREE-DIMENSIONAL DOMAIN SWAPPING AND ITS RELEVANCE TO CONFORMATIONAL DISEASES

MARIUSZ JASKOLSKI

Department of Crystallography, Faculty of Chemistry, A. Mickiewicz University and Center for Biocrystallographic Research, Polish Academy of Sciences, Grunwaldzka 6, 60–780 Poznan, Poland

Abstract: When a protein undergoes oligomerization via three-dimensional (3D) domain swapping, its molecules exchange secondary structure elements recreating the monomeric fold in an aberrant way, i.e., from chain segments belonging to different molecules. There is a hypothetical possibility that if this process took place in an open-ended, rather than reciprocal, fashion it could lead to the formation of pathological amyloid fibrils, which are associated with several conformational disorders. 3D domain swapping and disease-causing amyloid aggregation have been reported for many proteins, but human cystatin C (HCC) and the prion protein (PrP) are the only examples for which both phenomena have been observed.

Keywords: domain swapping; misfolding; amyloid fibril; amyloidoses; prion; conformational disorders; protein oligomerization; protein aggregation; cross-β structure; cystatin C.

1. Introduction

For a long time, it appeared that the so-called Anfinsen postulate, popularized in the phrase "One sequence – one fold," was a universal dogma, simplifying the intricate workings of molecular biology; at least one could be certain that in the end the genetic information was translated to unique, cast-iron three-dimensional (3D) structure. This simple picture is no longer true. There are protein sequences that can fold in more than one way; there are even proteins that seem to be unstructured altogether. Among the cases of proteins with multiple folds, 3D domain swapping, which is a mechanism of protein oligomerization, is emerging as a phenomenon with both deleterious consequences and beneficial effects for protein evolution and function.

R.J. Read and J.L. Sussman (eds.): Evolving Methods for Macromolecular Crystallography, 145–163
© 2007. *Springer*

2. The phenomenon of 3D domain swapping

The phenomenon known today as "3D domain swapping" had been predicted over four decades ago based on ingenious, and now classic experiments with activity recovery in dimers of ribonuclease (RNase) A with partly knocked-out active sites [1, 2]. Later, Piccoli et al. [3] and Mazzarella et al. [4] discovered the intertwined nature of bovine seminal RNase (BS-RNase) dimers. However, full recognition of the potential of this phenomenon and the introduction of the term "3D domain swapping" to scientific literature are attributed to Eisenberg and coworkers who, using X-ray crystallography, discovered this mechanism in the dimerization of diphtheria toxin [5]. 3D domain swapping refers to exchange of identical structural elements, referred to somewhat incorrectly as "3D domains," by two (or more) protein subunits. Those "domains" can be short segments or secondary structure elements, or, indeed, large and complete functional domains. In a 3D domain-swapped oligomer, a structural element of one subunit takes the place of the identical structural element of another, subunit and vice versa, leading to the recreation of the monomeric fold ("folding unit") but in an aberrant way, from chain segments contributed by different subunits (Figure 1). In a protein capable of domain swapping, there is an obligatory flexible hinge region, usually (but not always) a loop, whose conformational change allows the molecule to partially unfold and then find another similarly open monomer. The main force for adhesion between the components of a domain-swapped oligomer resides in the "closed interface" between the swapped domains, which recreates the structure and interactions of the monomer. It is a powerful factor in the structure of the oligomer as it has evolved to provide stability of the monomeric molecule. The oligomeric species, however, also has a new or "open" interface between its components that is not found in the monomeric form. If the oligomer is to be more stable than the monomers, the extra stabilization energy must come from the interactions in the open interface. It should be noted that, at best, only part of the energetic gain in the open interface will contribute to the stability of the oligomer because the rest of it will need to compensate for the entropic factor (loss of translational and rotational freedom), which always favors the monomer.

To date, about 50 proteins have been reported to undergo 3D domain swapping [6]. The list includes mostly dimers, but trimers and higher-order oligomers are also known. The most puzzling feature of this list is the lack of any common characteristics as it includes proteins of widely diversified structure, function, and origin. Taken rigorously, 3D domain swapping requires the same amino acid sequence to exist in both the monomeric form and as an intertwined oligomer. In practice, the usage of the term is more liberal, for instance, tolerating some sequence differences provided the folding

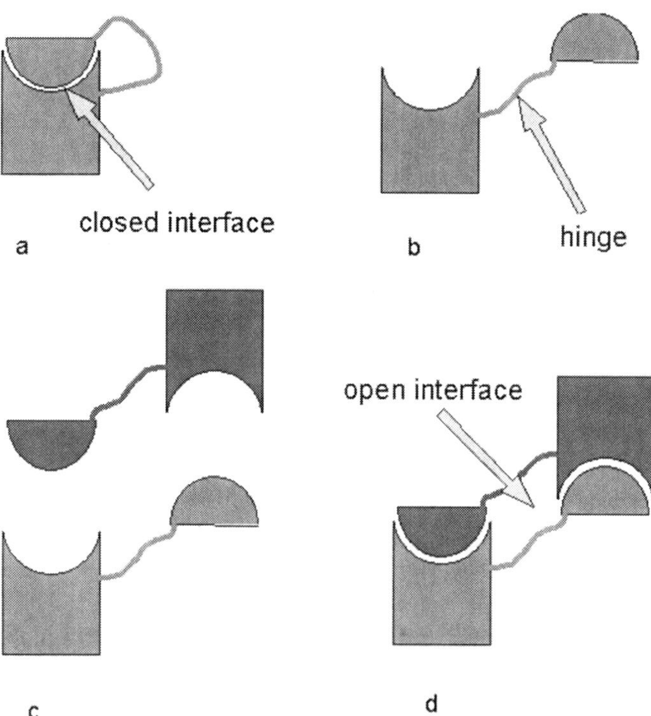

Figure 1. Cartoon illustration of dimer formation via 3D domain swapping. The compact globular fold (a) is partially unfolded (b) through a conformational change at a flexible hinge region. The unfolding temporarily disrupts and exposes the closed interface, i.e., the contact area between the two domains. If sufficiently long-lived, and if present in sufficiently high concentration (c), the unfolded chains will mutually recognize their complementary interfaces and will recreate those contacts in a symmetrical, dimeric fashion (d). Through the closed interfaces, two monomeric folds are reconstructed. However, the dimer is not a simple sum of two monomeric molecules. The hinge regions in the new conformation form a new intermolecular interface that was not present in the monomer. This is the open interface.

pattern is the same, or the term could even be applied when the existence of the monomeric species is not certain at all.

The classic example in the field of 3D domain swapping is bovine pancreatic RNase A. RNases are a large family of monomeric proteins for which no function other than RNA hydrolysis is known. The only exception is BS-RNase, which is naturally dimeric and possesses in this form interesting additional properties like allosteric regulation of the two active sites, cytotoxicity, and antiviral activity. BS-RNase is particularly intriguing because, genetically, it is present as a (defective) pseudogene in all ruminants except in ox (*Bos taurus*) and water buffalo (*Bubalus bubalis*), where the gene is functional. In the latter case, however, in variance with the high levels of BS-RNase in bovine seminal plasma, the protein is never expressed. Although the BS-RNase dimer has a covalent nature owing to two intermolecular disulfide bridges formed between uniquely placed cysteine residues, it exists in an equilibrium with about two-thirds of the molecules having an additional quaternary connection through an exchange of the N-terminal helix [4]. The special case of BS-RNase dimers interconnected through both covalent (S–S) and quaternary (domain–swap) interactions is of importance for the discussion by which of those two mechanisms ("covalent first" or "swap first") these dimers arose [7]. If the priming event was domain exchange triggered by a mutation or environmental change, the phenomenon of 3D domain swapping would be not only a structural curiosity, but also a powerful mechanism for rapid evolution of proteins from monomeric towards oligomeric forms with new biological properties.

Inspired by this observation, researches have tried to characterize domain-swapped dimers of RNase A molecules, whose naturally monomeric structure had been established by Wlodawer with very high accuracy [8, 9]. Dimerization of RNase A (by lyophilization from acetic acid) had been achieved in the early experiments of Crestfield [1], but the crystallographic model establishing the structure of the dimer with N-terminal domain swap was published only in 1998 [10]. Surprisingly, this structure, although sharing the closed interface with dimeric BS-RNase (and, obviously, with monomeric RNase A), had a different open interface and thus a different overall shape. More recent crystallographic studies have established that RNase A can also oligomerize via exchange of the C-terminal β-strand, leading to the formation of dimers [11] or cyclic trimers [12]. Even more interestingly, swapping of the N- and C-terminal segments by a central RNase A molecule with two neighbors would create an open-ended trimer, whose propagation could lead to infinite linear aggregation. Precise atomic structure of the linear trimers has not been established, but their existence has been confirmed beyond doubt by size determination and their dissociation patterns [12].

The quickly accumulating evidence about protein oligomerization via 3D domain swapping is a strong signal that existence of alternative protein conformations must be taken as a real possibility.

3. Human cystatin C, an example

The list of structurally characterized cases of 3D domain swapping includes two amyloidogenic proteins associated with human diseases: the prion protein (PrP) [13] and human cystatin C (HCC) [14]. The 3D domain-swapped PrP dimer can arise only after disruption of an intramolecular S–S bridge, which is then recreated (in two copies) between the subunits. Because of this observation, the suggestion that amyloidogenic aggregation of the PrP could involve 3D domain swapping was at first regarded with skepticism [15]. However, later, in a series of ingenious redox experiments at controlled denaturing conditions, Lee and Eisenberg [16] showed that recombinant monomeric PrP can be converted not only into dimers, but also into amyloid-like fibrils, which can then be used to seed the fibril formation of fresh material.

HCC is involved in two types of amyloid disorders: in hereditary cystatin C amyloid angiopathy (HCCAA), in which an L68Q mutant of the protein occurring naturally in an Icelandic subpopulation is deposited as amyloid causing brain hemorrhage and death in early adulthood [17], and in disorders involving deposition of amyloid β fibrils with wild-type cystatin C as a coprecipitant [18]. The single polypeptide chain of HCC is comprised of 120 amino acid residues and contains two disulfide bridges in the C-terminal part of the molecule [19]. HCC is present at high levels in all body fluids, in particular, in the cerebrospinal fluid, where it acts as one of the most important extracellular and transcellular inhibitors of papain-like cysteine proteases [20]. For the noncovalent inhibition of those enzymes it uses an epitope consisting of three elements aligned at one edge of the molecule, namely the N-terminal peptide, and two hairpin loops, L1 and L2. These loops are elements of a large and curved five-stranded antiparallel β-sheet which together with a long perpendicular α-helix forms a β-grip motif. The connectivity within this topology is (N)–β1–α–β2–L1–β3–AS–β4–L2–β5–(C), where AS is a broad "appending structure" positioned at the opposite ("back-side") end of the β-sheet relative to the N/L1/L2 edge. This monomeric fold was deduced from the structure of the related chicken cystatin [21] because crystallization experiments of HCC invariably lead to 3D domain-swapped dimers despite the use of size-exclusion chromatography in the final purification step [14, 22, 23]. The dimerization is promoted by incubation at high protein concentration and/or by mildly denaturing conditions or elevated temperature [24]. Dimers of the L68Q mutant are formed spontaneously and are also found in blood plasma and cerebrospinal fluid of HCCAA patients [25, 26].

The swapped element in HCC dimers consists of the α-helix and the two flanking strands, β1 and β2 (Figure 2). In the dimers, the monomeric fold of chicken cystatin is reconstructed with high fidelity. A single polypeptide chain "extracted" from the dimeric context demonstrates that the monomeric molecule underwent partial unfolding through an opening movement of the inhibitory loop L1. The absence of this loop in the dimeric structure is consistent with the observation that dimeric HCC has absolutely no inhibitory effect on papain-type proteases [25, 27]. The L1 hinge movement creates an unnatural conformation, necessitating domain complementation

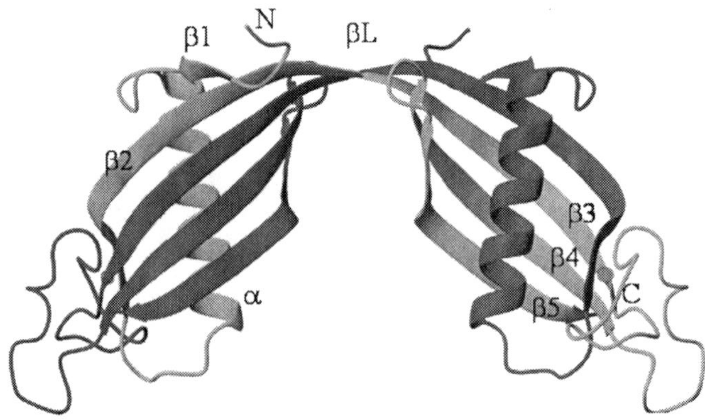

Figure 2. Twofold symmetric 3D domain swapped dimer of human cystatin C. The β-strands forming the antiparallel β-sheet in each domain are numbered from β1 to β5. They form a grip around the α-helix, which runs across the face of the β-sheet. Strands β2 traverse from one domain to the other (where they continue as β3) through the βL segment (open interface), which is created by opening of loop L1 present in the monomer. The dots mark the position of residue L68, which in the highly amloidogenic Icelandic mutant is replaced by glutamine.

with another unfolded chain, in order to bury the exposed surfaces that are not evolved to interact with water. In its new conformation, the L1 segment becomes part of a long β-ribbon running from the beginning of strand β2 to the end of strand β3. In addition to the monomer-type closed interface, the dimer also contains a new open interface, formed through β-sheet interactions in the L1 region (βL). This leads to the creation of an unusually long contiguous antiparallel β-sheet formed by two copies of the β2–βL–β3 ribbon, which cross from one domain to the other with as many as 34 main–chain ... main–chain hydrogen bonds and extra hydrogen bonds involving side chains.

Disruption of the monomeric fold requires both separation of the α-helix from the grip of the β-sheet as well as tearing of one of the seams of the β-sheet itself (β2–β3). Conversely, the closed interface reconstructed in the dimer is cemented by the same interactions. The two disulfide bridges introducing rigidity into the fold are both present in the C-terminal domain and in consequence do not interfere with the domain swapping process, but help to maintain the integrity of the C-terminal domain during the transition period when the protein must undergo partial unfolding.

3D Domain swapped full-length HCC has been characterized in two crystal forms: cubic [14] and tetragonal (unpublished results). In both cases, the swapped elements (and the closed interface) are the same. However, the conformation of the hinge region at the open interface is very different

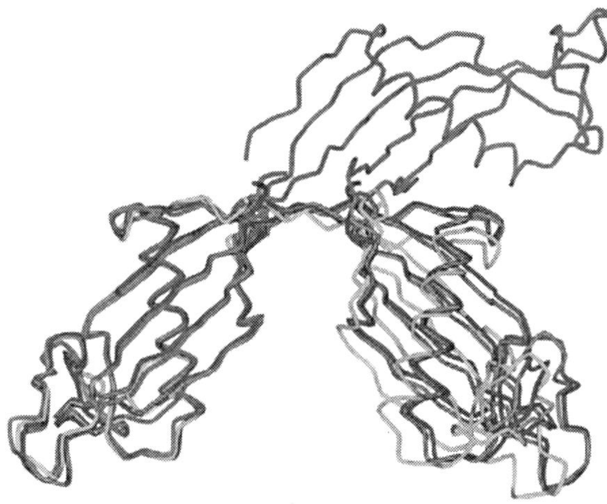

Figure 3. Superposition of the dimeric HCC structures emphasizing the plasticity of the molecules originating from the flexibility of the hinge element at the open interface. The crystallographic dimer of the cubic crystal form of HCC (PDB code 1G96), dimers CD and EF of the N-truncated variant (THCC, PDB code 1R4C), and the noncrystallographic dimer of the tetragonal crystal form of HCC (PDB code 1TIJ) are shown in shades of gray. The superpositions were calculated using the Cα-atoms from only one half (left-hand side) of the molecules. The differences in the dimer geometry are therefore emphasized in the right-hand side domain.

resulting in a different overall molecular shape (Figure 3). This is in line with the findings reported for RNase A that for a given protein there may be several possibilities to form domain-swapped oligomers. With regard to HCC, it demonstrates the high degree of flexibility of the dimer and its structural adaptability to the molecular environment, such as can be found in a crystal or in the amyloid fibril.

Protein extracted from cystatin C amyloid is shortened at the N-terminus [28]. Although it is not certain if the truncation precedes aggregation or is connected with fibril processing or the separation procedure, it was of interest to see if N-truncated HCC (THCC) is also capable of 3D domain swapping. The crystal structure of a recombinant protein with an analogous 10-amino-acid N-terminal truncation shows eight polypeptide chains paired into 3D domain-swapped dimers in a way that is very similar to that observed in the cubic crystal structure of the full-length protein (Figure 3) [23].

Position 68 in the amino acid sequence of HCC is the only site at which the wild-type protein (L68) and the highly amyloidogenic variant (Q68) associated with the Icelandic-type cerebral angiopathy (HCCAA) differ. All crystal structures of HCC (full-length and N-truncated) reveal that leucine 68, located on the central β3-strand of the β-sheet on its inner face (Figure 2), is buried by the α-helix. In the 3D domain-swapped dimers, the β3- and α-elements are contributed by different subunits, but since this interaction is part of the closed interface, identical contacts must exist in monomeric cystatin C. In the hydrophobic core of the protein, L68 occupies a pocket formed by the surrounding residues on the β-sheet and the hydrophobic face of the helix. Replacement of the hydrophobic side chain of L68 by the longer glutamine side chain in the pathological variant not only makes those contacts prohibitively close, but also places the mutated hydrophilic chain in hydrophobic environment. This must definitely destabilize the molecular α–β interface and lead to repulsive interactions expelling the α-helix (together with strand β2) from the compact molecular core. Such a movement is possible through a conformational change of the L1 hinge. This explains the increased dynamic properties of the L68Q mutant compared with wild-type cystatin C as observed by nuclear magnetic resonance (NMR) spectroscopy [29, 30]. Under the assumption that the refolded dimer recreates the topology of monomeric cystatin C, those destabilizing effects would be similar in both cases. However, the dimeric structure may be more resistant to disruption because of the extra stabilization contributed by the β-interactions in the linker region. The hydrophilic L68Q substitution at the α–β interface is also expected to reduce the unfavorable solvent contacts when this surface is exposed upon unfolding, thus lowering the barrier for monomer→dimer transitions and making the latter form kinetically more accessible. The above discussion of the effect of the L68Q substitution on HCC dimerization is supported by the observation that the mutated variant forms dimers in *in vitro* experiments and in blood plasma much more easily than wild-type cystatin C [24, 26].

Although it is generally accepted that the hinge sequence is the best region for controlling 3D domain swapping [10], mutations outside of the hinge have also to be considered [31]. In the case of HCC, the L68Q substitution decreases the energy necessary for the transition from the monomeric to the dimeric form by destabilizing the monomer (higher energy) and by lowering the activation energy of the partially unfolded state (less unfavorable interactions with solvent in the open conformation). This is in line with the remark by Perutz [32] that mutations of "internal" residues may result in a loss in free energy of stabilization which, even if small, might lead to a disruptive "loosening" of the native structure.

HCC does not represent a strict *bona fide* 3D domain swapping case as defined by Eisenberg [33] because the structure of the monomeric protein is not precisely known. However, it is certain that such monomers do exist and one can be quite confident that their structure closely resembles that of the chicken homolog.

4. Amyloid criteria and structure

The term "amyloid" has an interesting history [34–36]. First descriptions of foreign-deposit-laden postmortem tissues and organs appear in the seventeenth century. In scientific literature, the term "amyloid," which indicated carbohydrate because of iodine staining, was used for the first time by Virchow in 1854, a few years before Friedreich and Kekule demonstrated that amyloid deposits were of predominantly proteinaceous character. To add to the confusion, it has to be admitted that *corpora amylacea*, on the observation of which Virchow coined his term, have been recently found to indeed consist primarily of polysaccharides, and as such are not amyloids in our present use of the term. It refers to a less picturesque object, a pathological proteinaceous substance deposited extracellularly or intracellularly in tissues, having fibrous microscopic morphology, clinically leading to tissue damage, and typically connected with lethal diseases.

Amyloid deposits are formed from proteins that are otherwise soluble in their physiological role. The list of proteins with confirmed amyloidogenic properties has grown in recent years to include more than 20 cases. In addition to such familiar examples as the Alzheimer amyloid β protein or the PrP, there is a whole range of proteins with a diverse spectrum of biological functions, for example, lysozyme, insulin, or transthyretin. The discovery that amyloid formation can be reproduced *in vitro* and that it is not restricted to a limited number of protein sequences associated with diseases [37] has also significantly enlarged the field of study. It has been proposed that proteins may contain "chameleon" sequences, equally unstable in the α and β conformation, that would destabilize protein structure and promote unfolding [32, 38, 39]. If such labile chameleon sequences were common, the potential for conformational aberrations would be much higher than is currently believed. Another intriguing point about amyloid is that, while it is often related to or triggered by a mutated variant of a normally stable protein, its formation can also occur in the unmutated form. This is observed, for instance, for HCC and the PrP, in which case the misfolded or conformationally defective form is considered to be the transmissible pathogenic agent.

Production of artificial amyloid fibrils can be illustrated with the experiments using HCC, which can be converted into fibrillar form using mildly denaturing conditions, for instance, incubation at 3 mg/ml concentration in 10 mM glycine buffer (pH 2.0) at 48 °C under constant stirring (Figure 4). Interestingly, mildly denaturing conditions also induce dimerization of wild-type HCC [24].

There are three basic criteria that must be met by amyloid deposits, connected with their tinctorial, morphological, and structural characteristics [35].

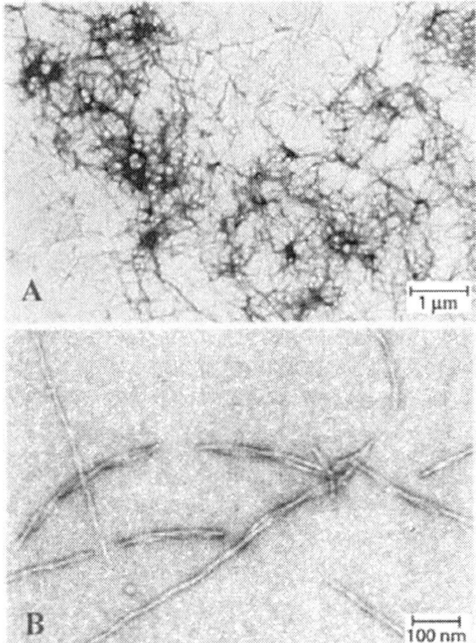

Figure 4. Electron micrographs of fibrils formed from wt cystatin C. *A* and *B* show, at different magnifications, fibrils produced from a solution of wt cystatin C (3 mg/ml) in 10 mM
glycine buffer, pH 2.0, stirred at 48°C for 7 days.

Amyloids have specific tinctorial properties, i.e., are stained when treated
with organic dyes, such as thioflavine T or Congo Red. In the test using
Congo Red, amyloids are stained to give a characteristic apple green bire-
fringence [40] when viewed in polarized light. Secondly, electron micrographs
of amyloid deposits show them to be composed of uniform and straight
fibers with ~100 Å diameter [41]. Thirdly, X-ray diffraction patterns of oriented amyloid fibrils show them to have ordered, repeating structure, consistent with the so-called cross-β structure [42, 43], in which extended
polypeptide chains in β-conformation are perpendicular to the fiber axis, and
form helically twisted β-sheets that are parallel to the fiber axis. Those
molecular features have been deduced from the observation of a strong
meridional reflection with a d spacing of about 4.7 Å, corresponding to the
repeating distance between the β-strands, and from several orders of a
$d \approx 115$ Å reflection, representing the repeat distance (pitch) of the helically
twisted β-sheet. Additionally, an equatorial reflection with a d spacing of
about 10 Å is interpreted to represent the distance between the β-sheets in
the fibril [44]. In particular, synchrotron X-ray studies have suggested that
the core of the transthyretin amyloid fibril is a continuous β-sheet helix [45].

The degree of similarity in the diffraction patterns of amyloid fibers produced from very different polypeptides is indicative of a common core structure, which must be assumed in the fibril regardless of the soluble-form properties of the constituent protein and despite the known, large differences in folding of the precursor proteins.

Recently, Wille et al. [46] proposed a model for the amyloid fibril cross-β structure that is different from the commonly accepted β-sheet helix illustrated in Figure 5. In their model, based on electron microscopic studies of two-dimensional (2D) crystals of the scrapie variant of human prion protein (PrP^{Sc}), the β-strands form a parallel β-helix (either right- or left-handed), similar to that discovered in pectate lyase [47].

5. 3D Domain swapping and amyloid aggregation

Although 3D domain swapping has been proposed as a mechanism of amyloid fibril formation [5, 48, 49], there are only a limited number of experimental studies linking directly 3D domain swapping and amyloid aggregation. In one study, Ogihara et al. [50] engineered two 3D domain-swapped derivatives of two helical protein scaffolds, designed to undergo either 3D domain swapping dimerization or 3D domain swapping multimeric fibrous assembly. The predicted assemblies were detected by a variety of physicochemical methods and although the structure of the fibrils that formed in the second case was not established at the molecular level, all other evidence points to 3D domain swapping as the mechanism through which those fibrils were formed. It has to be admitted, however, that, as designed, those structures were predominantly helical while the generally accepted molecular architecture of amyloid fibrils, of both natural and artificial origin, is essentially of β-character. More recently, Lee and Eisenberg [16] using controlled denaturing and redox conditions showed that the S–S bridge present in the molecule of recombinant PrP can be reestablished on 3D domain swapping not only in dimeric context [13], but also in amyloid fibrils.

Another experimental link between 3D domain swapping and amyloid formation is provided by mutagenesis studies of HCC aimed at stabilizing the monomeric fold against 3D domain swapping by the introduction of new disulfide bridges. Two variants of both wild-type and L68Q cystatin C were produced [24]. One of the disulfide bridges was designed to link the α-helix and the β-sheet, and the other to connect strands $\beta2$ and $\beta3$, i.e., in each case a covalent link was introduced between the secondary structure elements that must separate in the act of 3D domain swapping (Figure 6). The two monomeric domain-stabilized forms of both wild-type and L68Q cystatin C were then incubated at conditions known to convert virtually all monomers

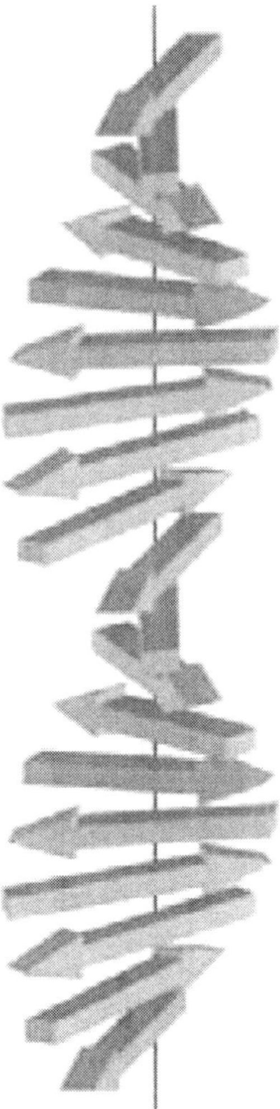

Figure 5. The "cross-β structure" revealed by diffraction studies of amyloid fibrils is usually explained by a β-sheet helix, with individual β-strands perpendicular to the helix axis, and the β-sheet face parallel to the helix axis. The pitch of the β-sheet helix has been estimated at about 115 Å, corresponding to 24 β-chains with a repeat distance of about 4.7 Å [35]. For clarity, the number of β-rungs per repeat of the β-sheet ladder in the diagram is arbitrary, less than found experimentally.

of cystatin C into dimers and it was observed that no dimers were formed by any of the stabilized proteins. The formation of amyloid fibrils by the stabilized monomers of wild-type cystatin C was also tested, by incubation at conditions known to produce large amounts of amyloid fibrils. Amyloid fibril formation was reduced by about 80% for both forms of the stabilized protein.

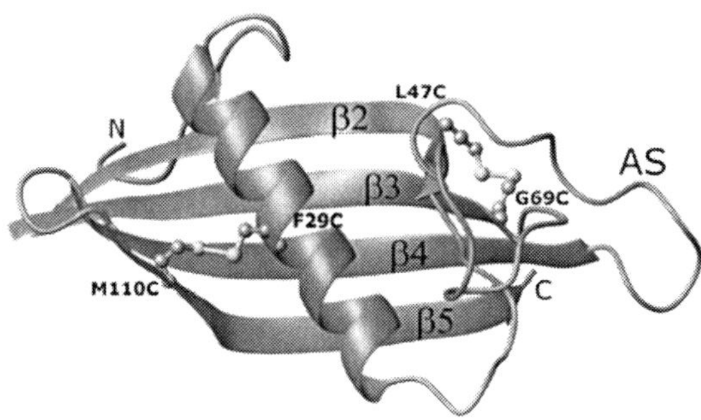

Figure 6. Positioning of S–S bridges introduced to stabilize the monomeric fold of cystatin C. One pair of Cys mutations introduces a new disulfide bond linking the strands β2 and β3 of the β-sheet. The other pair of mutations creates a link between the α-helix and strand β5. In the domain-swapped dimer of cystatin C, strand β2 and the α-helix are contributed by one of the monomers, and strands β3–β5 by the other.

Although this significant reduction is in line with the notion that 3D domain swapping might be important for amyloid fibril formation, the fact that some residual amounts of fibrils are nevertheless formed, requires further consideration. One explanation could be that 3D domain swapping is not absolutely necessary for amyloid aggregation. Another, more likely possibility is that under the harsh conditions of the experiment some of the stabilizing disulfide bridges are disrupted, sending the molecules on the fibril formation pathway.

Although the production of cystatin C S–S stabilized variants is very useful for demonstrating that prevention of 3D domain swapping has indeed the power to inhibit formation of dimers and amyloid fibrils, it is obvious that if treatment strategies are to be developed to prevent amyloidogenesis, exogenous agents stabilizing the monomeric form must be sought. It is therefore of interest that catalytic amounts of a monoclonal antibody raised against wild-type cystatin C can be used to inhibit the dimerization process of both wild-type and L68Q cystatin C [24]. This is reminiscent of the discovery that antibodies to surface epitopes of PrP^C can inhibit the generation of the amyloidogenic species PrP^{Sc} and thereby interfere with prion biogenesis [51]. It is not clear why the antibody is so efficient in preventing dimerization even at very low molar ratio. One possibility is that it reacts with some intermediate form on the dimerization pathway.

As mentioned above, cysteine proteases of the papain family can only interact with monomeric cystatin C and might, therefore, also be potential agents for the stabilization of this form. Indeed, recent results show that carboxymethylpapain, an active site-alkylated derivative of the enzyme, inhibits the formation of dimers from both wild-type and L68Q cystatin C, analogously to the monoclonal antibody [24].

An interesting connection between 3D domain swapping and amyloid structure is provided by the crystal structure of tetragonal HCC (unbublished results). Here, 3D domain-swapped dimers formed as described above, pack in the crystal lattice via β-sheet interactions at the free edges of their molecular β-sheets, extending them into intermolecular context. In consequence, a supramolecular crystal structure is generated, with all the β-strands of the domain-swapped dimers perpendicular to a common direction, an arrangement that is reminiscent of the cross-β structure of amyloid fibrils.

There are striking parallels between the phenomena of 3D domain swapping and amyloid fibril formation and they provide strong circumstantial evidence that 3D domain swapping might be involved in amyloidogenesis. For instance, both processes are highly selective with respect to their building blocks, but occur in all kinds of proteins, regardless of structure, function, or origin. They are facilitated by mildly denaturing conditions but the new, aberrant structure seems to be generated with very high precision. Both require conformational transformation and are initiated by crossing high barriers of activation. Finally, in the body fluids of patients with the trait for HCCAA, dimers of the L68Q mutant are detected [26], which by all available evidence are identical with the crystallographic 3D domain swapped dimers of cystatin C.

But even if we agree that 3D domain swapping might be indeed involved in the formation of amyloid fibrils, several different scenarios may be envisaged.

1. The aggregating protein could be capable of swapping different domains at both termini. Such a possibility has been found for RNase A [11, 12], but in general, it does not appear to be very common. In this scenario, the growing polymer would not have "sticky ends," or incomplete domains necessitating complementation through another domain swapping event; the ends would be protected and inert from the point of view of aggregation until an unfolding event exposed the domains for swapping again. Therefore, kinetically and thermodynamically, the growth phase would not be very different from the nucleation phase.

2. The domain that is known to undergo closed-ended, mutual swapping could, in principle, undergo infinite open-ended swapping, leading to linear polymerization. In this scenario, the initial and late stages of aggregation would be markedly different. During the growth phase, there would be a permanent presence of unsatisfied open domains at both ends of each growing fibril, providing sticky ends for continuing growth. Prior

to the formation of stable seeds, however, the unfolded monomers, presumed to be in equilibrium with the folded form, could always refold back as monomers or closed-ended circular oligomers, thus eliminating the points of attachment (unsatisfied domains) from the system. Here, formation of closed oligomers (e.g., dimers) could be a dead end on the oligomerization pathway, and in a way could prevent, or slow down, the process of amyloid fibril formation [52].

3. It is conceivable that in amyloidogenic aggregation two different mechanisms could be in operation. The building blocks of the amyloid fibril could be formed as closed-ended domain-swapped oligomers, which would aggregate further using a different mechanism, for example, intermolecular β-type association, as in the tetragonal crystal structure of HCC. In this scenario, the accumulation of closed-ended oligomers would not be viewed as a suicidal trap on the polymerization pathway [52], but would be a required initial step providing "substrates" for the final polymerization step.

6. 3D Domain swapping and protein evolution

By linking 3D domain swapping with folding and conformational aberrations of amyloidogenic proteins one tends to emphasize the dark side of this phenomenon. However, an increasing number of reports suggest that 3D domain swapping may also have a positive role in protein evolution and function.

One obvious advantage of multiplicating the active sites in 3D domain-swapped enzymes would be the possibility of allosteric regulation, as is for example, observed in BS-RNase [53]. BS-RNase is particularly interesting, because in variance with other RNases, in its dimeric form it can cleave double-stranded RNA. Other examples of functional regulation of proteins have also been reported. For instance, glyoxalase I exists as an active 3D domain-swapped dimer and as a less active metastable monomer, and this conversion is controlled by glutathione [54, 55]. It has been also shown that 3D domain swapping may be involved in controlling protein–protein interactions, for example, through the formation of oligomeric receptor molecules [56] or in the creation of more stable viral capsids [57]. All the above roles of 3D domain swapping are connected with some more general "economic" advantages: reduction of genome size with simultaneous increase of stability and complexity [58].

When 3D domain swapping, triggered, for example, by an environmental change, leads to gain of new function, this new function could be quickly "fixed" by stabilizing mutations. In this way, 3D domain swapping could be a "fast track" for protein evolution. It has been postulated that some present-day side-by-side dimers might have evolved via a priming domain exchange act which was later lost as a consequence of stabilizing mutations [59]. An opposite evolutionary role has been demonstrated by Chirgadze et al. [60], who showed that 3D domain swapping can rescue protein function after destabilizing mutation.

It is argued sometimes that the cases of 3D domain swapping observed by X-ray crystallography do not reflect in vivo situations, but are artifacts resulting from prolonged incubation at high protein concentration and (possibly) nonphysiological pH [58]. While this may be true in some of the cases, it is difficult to accept that all the diverse crystallographic domain-swapped oligomers could be artifacts of crystallization. In addition, 3D domain swapping has been also observed by NMR spectroscopy [61] and deduced from other physicochemical measurements, e.g., size determination. We may thus conclude that protein oligomerization (and possibly aggregation) by 3D domain swapping is not uncommon, and that its consequences may be both beneficial for protein function and evolution as well as deleterious, when the aberrant conformation leads to dysfunction and disease. On the other hand, proteins may have protective mechanisms to avoid harmful aggregation arising from optimizing function. For instance, the edges of their open β-sheets may have evolved in most cases to be less favorable for β-type aggregation [62].

Acknowledgments

This work was supported, in part, by a grant from the State Committee for Scientific Research (4 T09A 039 25), by a subsidy from the Foundation for Polish Science, and by a Faculty Scholar fellowship from the National Cancer Institute (USA).

References

1. Crestfield, A.M., Stein, W.H., and Moore, S. (1962) On the aggregation of bovine pancreatic ribonuclease. *Archives of Biochemistry and Biophysics*, Suppl. 1: 217–222.
2. Crestfield, A.M., Stein, W.H., and Moore, S. (1963) Properties and conformation of the histidine residues at the active site of ribonuclease. *Journal of Biological Chemistry*, **238**: 2421–2428.
3. Piccoli, R., Tamburrini, M., Piccialli, G., Di Donato, A., Parente, A., and D'Alessio, G. (1992) The dual-mode quaternary structure of seminal RNase. *Proceedings of the National Academy of Sciences of the USA*, **89**: 1870–1874.
4. Mazzarella, L., Capasso, S., Demasi, D., Di Lorenzo, G., Mattia, C.A., and Zagari, A. (1993) Bovine seminal ribonuclease: structure at 1.9 Å resolution. *Acta Crystallographica*, **D49**: 389–402.

5. Bennett, M.J., Choe, S., and Eisenberg, D. (1994) Domain swapping: entangling alliances between proteins. *Proceedings of the National Academy of Sciences of the USA*, **91**: 3127–3131.

6. Liu, Y. and Eisenberg, D. (2002) 3D domain swapping: as domains continue to swap. *Protein Science*, **11**: 1285–1299.

7. D'Alessio, G. (1999) Evolution of oligomeric proteins. The unusual case of a dimeric ribonuclease. *European Journal of Biochemistry*, **266**: 699–708.

8. Wlodawer, A., Bott, R., and Sjolin, L. (1982) The refined crystal structure of ribonuclease A at 2.0 Å resolution. *Journal of Biological Chemistry*, **257**: 1325–1332.

9. Wlodawer, A., Svensson, L.A., Sjolin, L., and Gilliland, G.L. (1988) Structure of phosphate-free ribonuclease A refined at 1.26 Å. *Biochemistry*, **27**: 2705–2717.

10. Liu, Y., Hart, P.J., Schlunegger, M.P., and Eisenberg, D. (1998) The crystal structure of a 3D domain-swapped dimer of RNase A at a 2.1-A resolution. *Proceedings of the National Academy of Sciences of the USA*, **95**: 3437–3442.

11. Liu, Y., Gotte, G., Libonati, M., and Eisenberg, D. (2001) A domain-swapped RNase A dimer with implications for amyloid formation. *Nature Structural and Molecular Biology*, **8**: 211–214.

12. Liu, Y., Gotte, G., Libonati, M., and Eisenberg, D. (2002) Structures of the two 3D domain-swapped RNase A trimers. *Protein Science*, **11**: 371–380.

13. Knaus, K.J., Morillas, M., Swietnicki, W., Malone, M., Surewicz, W.K., and Yee, V.C. (2001) Crystal structure of the human prion protein reveals a mechanism for oligomerization. *Nature Structural and Molecular Biology*, **8**: 770–774.

14. Janowski, R., Kozak, M., Jankowska, E., Grzonka, Z., Grubb, A., Abrahamson, M., and Jaskolski, M. (2001) Human cystatin C, an amyloidogenic protein, dimerizes through three-dimensional domain swapping. *Nature Structural and Molecular Biology*, **8**: 316–320.

15. Nicholson, E.M., Mo, H., Prusiner, S.B., Cohen, F.E., and Marqusee, S. (2002) Differences between the prion protein and its homolog Doppel: a partially structured state with implications for scrapie formation. *Journal of Molecular Biology*, **316**: 807–815.

16. Lee, S. and Eisenberg, D. (2003) Seeded conversion of recombinant prion protein to a disulfide-bonded oligomer by a reduction–oxidation process. *Nature Structural and Molecular Biology*, **10**: 725–730.

17. Olafsson, I. and Grubb, A. (2000) Hereditary cystatin C amyloid angiopathy. *Amyloid*, **7**: 70–79.

18. Maruyama, K., Ikeda, S., Ishihara, T., Allsop, D., and Yanagisawa, N. (1990) Immunohistochemical characterization of cerebrovascular amyloid in 46 autopsied cases using antibodies to beta protein and cystatin C. *Stroke*, **21**: 397–403.

19. Grubb, A. and Lofberg, H. (1982) Human gamma-trace, a basic microprotein: amino acid sequence and presence in the adenohypophysis. *Proceedings of the National Academy of Sciences of the USA*, **79**: 3024–3027.

20. Abrahamson, M., Barrett, A.J., Salvesen, G., and Grubb, A. (1986) Isolation of six cysteine proteinase inhibitors from human urine. Their physicochemical and enzyme kinetic properties and concentrations in biological fluids. *Journal of Biological Chemistry*, **261**: 11282–11289.

21. Bode, W., Engh, R., Musil, D., Thiele, U., Huber, R., Karshikov, A., Brzin, J., Kos, J., and Turk, V. (1988) The 2.0 A X-ray crystal structure of chicken egg white cystatin and its possible mode of interaction with cysteine proteinases. *The EMBO Journal*, **7**: 2593–2599.

22. Kozak, M., Jankowska, E., Janowski, R., Grzonka, Z., Grubb, A., Alvarez, F.M., Abrahamson, M., and Jaskolski, M. (1999) Expression of a selenomethionyl derivative and preliminary crystallographic studies of human cystatin C. *Acta Crystallographica Section D: Biological crystallography*, **55**: 1939–1942.

23. Janowski, R., Abrahamson, M., Grubb, A., and Jaskolski, M. (2004) Domain swapping in N-truncated human cystatin C. *Journal of Molecular Biology*, **341**: 151–160.

24. Nilsson, M., Wang, X., Rodziewicz-Motowidlo, S., Janowski, R., Lindstrom, V., Onnerfjord, P., Westermark, G., Grzonka, Z., Jaskolski, M., and Grubb, A. (2004)

Prevention of domain swapping inhibits dimerization and amyloid fibril formation of cystatin C: use of engineered disulfide bridges, antibodies, and carboxymethylpapain to stabilize the monomeric form of cystatin C. *Journal of Biological Chemistry*, **279**: 24236–24245.

25. Abrahamson, M. and Grubb, A. (1994) Increased body temperature accelerates aggregation of the Leu-68—— > Gln mutant cystatin C, the amyloid-forming protein in hereditary cystatin C amyloid angiopathy. *Proceedings of the National Academy of Sciences of the USA*, **91**: 1416–1420.

26. Bjarnadottir, M., Nilsson, C., Lindstrom, V., Westman, A., Davidsson, P., Thormodsson, F., Blondal, H., Gudmundsson, G., and Grubb, A. (2001) The cerebral hemorrhage-producing cystatin C variant (L68Q) in extracellular fluids. *Amyloid*, **8**: 1–10.

27. Ekiel, I. and Abrahamson, M. (1996) Folding-related dimerization of human cystatin C. *Journal of Biological Chemistry*, **271**: 1314–1321.

28. Ghiso, J., Jensson, O., and Frangione, B. (1986) Amyloid fibrils in hereditary cerebral hemorrhage with amyloidosis of Icelandic type is a variant of gamma-trace basic protein (cystatin C). *Proceedings of the National Academy of Sciences of the USA*, **83**: 2974–2978.

29. Ekiel, I., Abrahamson, M., Fulton, D.B., Lindahl, P., Storer, A.C., Levadoux, W., Lafrance, M., Labelle, S., Pomerleau, Y., Groleau, D., LeSauteur, L., and Gehring, K. (1997) NMR structural studies of human cystatin C dimers and monomers. *Journal of Molecular Biology*, **271**: 266–277.

30. Gerhartz, B., Ekiel, I., and Abrahamson, M. (1998) Two stable unfolding intermediates of the disease-causing L68Q variant of human cystatin C. *Biochemistry*, **37**: 17309–17317.

31. Rousseau, F., Schymkowitz, J.W., and Itzhaki, L.S. (2003) The unfolding story of three-dimensional domain swapping. *Structure (Camb.)*, **11**: 243–251.

32. Perutz, M.F. (1997) Amyloid fibrils. Mutations make enzyme polymerize. *Nature*, **385**: 773, 775.

33. Schlunegger, M.P., Bennett, M.J., and Eisenberg, D. (1997) Oligomer formation by 3D domain swapping: a model for protein assembly and misassembly. *Advances in Protein Chemistry*, **50**: 61–122.

34. Cohen, A.S. (1986). General introduction and a brief history of the amyloid fibril. In *Amyloidosis*. Edited by Marrink, J. and Van Rijvijk, M.H. Dordrecht: Nijhoff, pp. 3–19.

35. Sunde, M. and Blake, C.C. (1998) From the globular to the fibrous state: protein structure and structural conversion in amyloid formation. *Quarterly Reviews of Biophysics*, **31**: 1–39.

36. Sipe, J.D. and Cohen, A.S. (2000) Review: history of the amyloid fibril. *Journal of Structural Biology*, **130**: 88–98.

37. Dobson, C.M. (1999) Protein misfolding, evolution and disease. *Trends in Biochemical Sciences*, **24**: 329–332.

38. Booth, D.R., Sunde, M., Bellotti, V., Robinson, C.V., Hutchinson, W.L., Fraser, P.E., Hawkins, P.N., Dobson, C.M., Radford, S.E., Blake, C.C., and Pepys, M.B. (1997) Instability, unfolding and aggregation of human lysozyme variants underlying amyloid fibrillogenesis. *Nature*, **385**: 787–793.

39. Minor, D.L. Jr. and Kim, P.S. (1996) Context-dependent secondary structure formation of a designed protein sequence. *Nature*, **380**: 730–734.

40. Glenner, G.G., Eanes, E.D., and Page, D.L. (1972) The relation of the properties of Congo red-stained amyloid fibrils to the -conformation. *Journal of Histochemistry and Cytochemistry*, **20**: 821–826.

41. Cohen, A.S., Shirahamata, T., and Skinner, M. (1982). *Electron Microscopy of Amyloid*, In Electron Microscopy of Proteins. Edited by Harris, J.R. New York: Academic Press, pp. 165–205.

42. Glenner, G.G. (1980) Amyloid deposits and amyloidosis: the beta-fibrilloses (second of two parts). *The New England Journal of Medicine*, **302**: 1333–1343.

43. Glenner, G.G. (1980) Amyloid deposits and amyloidosis. The beta-fibrilloses (first of two parts). *The New England Journal of Medicine*, **302**: 1283–1292.

44. Sunde, M., Serpell, L.C., Bartlam, M., Fraser, P.E., Pepys, M.B., and Blake, C.C. (1997) Common core structure of amyloid fibrils by synchrotron X-ray diffraction. *Journal of Molecular Biology*, **273**: 729–739.

45. Blake, C. and Serpell, L. (1996) Synchrotron X-ray studies suggest that the core of the transthyretin amyloid fibril is a continuous beta-sheet helix. *Structure*, **4**: 989–998.

46. Wille, H., Michelitsch, M.D., Guenebaut, V., Supattapone, S., Serban, A., Cohen, F.E., Agard, D.A., and Prusiner, S.B. (2002) Structural studies of the scrapie prion protein by electron crystallography. *Proceedings of the National Academy of Sciences of the USA*, **99**: 3563–3568.

47. Lietzke, S.E., Yoder, M.D., Keen, N.T., and Jurnak, F. (1994) The Three-dimensional structure of pectate lyase E, a plant virulence factor from *Erwinia chrysanthemi*. *Plant Physiology*, **106**: 849–862.

48. Klafki, H.W., Pick, A.I., Pardowitz, I., Cole, T., Awni, L.A., Barnikol, H.U., Mayer, F., Kratzin, H.D., and Hilschmann, N. (1993) Reduction of disulfide bonds in an amyloidogenic Bence Jones protein leads to formation of "amyloid-like" fibrils in vitro. *Biological Chemistry Hoppe-Seyler*, **374**: 1117–1122.

49. Cohen, F.E. and Prusiner, S.B. (1998) Pathologic conformations of prion proteins. *Annual Review of Biochemistry*, **67**: 793–819.

50. Ogihara, N.L., Ghirlanda, G., Bryson, J.W., Gingery, M., DeGrado, W.F., and Eisenberg, D. (2001) Design of three-dimensional domain-swapped dimers and fibrous oligomers. *Proceedings of the National Academy of Sciences of the USA*, **98**: 1404–1409.

51. White, A.R. and Hawke, S.H. (2003) Immunotherapy as a therapeutic treatment for neurodegenerative disorders. *Journal of Neurochemistry*, **87**: 801–808.

52. Jaskolski, M. (2001) 3D domain swapping, protein oligomerization, and amyloid formation. *Acta Biochimica Polonica*, **48**: 807–827.

53. Vitagliano, L., Adinolfi, S., Sica, F., Merlino, A., Zagari, A., and Mazzarella, L. (1999) A potential allosteric subsite generated by domain swapping in bovine seminal ribonuclease. *Journal of Molecular Biology*, **293**: 569–577.

54. Saint-Jean, A.P., Phillips, K.R., Creighton, D.J., and Stone, M.J. (1998) Active monomeric and dimeric forms of Pseudomonas putida glyoxalase I: evidence for 3D domain swapping. *Biochemistry*, **37**: 10345–10353.

55. SaintJean, A.P. and Creighton, D.J. (1999) Regulated 3D domain swapping of glyoxalase I. *FASEB Journal*, **13**: A1560.

56. Gouldson, P.R., Higgs, C., Smith, R.E., Dean, M.K., Gkoutos, G.V., and Reynolds, C.A. (2000) Dimerization and domain swapping in G-protein-coupled receptors: a computational study. *Neuropsychopharmacology*, **23**: S60–S77.

57. Qu, C., Liljas, L., Opalka, N., Brugidou, C., Yeager, M., Beachy, R.N., Fauquet, C.M., Johnson, J.E., and Lin, T. (2000) 3D domain swapping modulates the stability of members of an icosahedral virus group. *Structure with Folding and Design*, **8**: 1095–1103.

58. Marianayagam, N.J., Sunde, M., and Matthews, J.M. (2004) The power of two: protein dimerization in biology. *Trends in Biochemical Sciences*, **29**: 618–625.

59. Bennett, M.J. and Eisenberg, D. (2004) The evolving role of 3D domain swapping in proteins. *Structure (Camb.)*, **12**: 1339–1341.

60. Chirgadze, D.Y., Demydchuk, M., Becker, M., Moran, S., and Paoli, M. (2004) Snapshot of protein structure evolution reveals conservation of functional dimerization through intertwined folding. *Structure (Camb.)*, **12**: 1489–1494.

61. Staniforth, R.A., Giannini, S., Higgins, L.D., Conroy, M.J., Hounslow, A.M., Jerala, R., Craven, C.J., and Waltho, J.P. (2001) Three-dimensional domain swapping in the folded and molten-globule states of cystatins, an amyloid-forming structural superfamily. *The EMBO Journal*, **20**: 4774–4781.

62. Richardson, J.S. and Richardson, D.C. (2002) Natural beta-sheet proteins use negative design to avoid edge-to-edge aggregation. *Proceedings of the National Academy of Sciences of the USA*, **99**: 2754–2759.

STRUCTURAL BIOINFORMATICS: FROM PROTEIN STRUCTURE TO FUNCTION

JAMES D. WATSON, ADEL GOLOVIN,
ROMAN A. LASKOWSKI, KIM HENRICK,
AND JANET M. THORNTON

*EMBL Outstation Hinxton, European Bioinformatics
Institute, Wellcome Trust Genome Campus, Hinxton,
Cambridge, CB10 1SD, UK*

ANDRZEJ JOACHIMIAK

*Biosciences Division, Midwest Center for Structural
Genomics and Structural Biology Center, Argonne National
Laboratory, 9700 S Cass Ave. Argonne, IL 60439, USA*

ALED M. EDWARDS

*Structural Genomics Consortium, University of Toronto,
100 College Street, Toronto, Ontario, M5G 1L5, Canada*

Abstract: A major problem faced by structural biology today is the issue of function prediction. With the success of the various Structural genomics initiatives and advances in crystallography, proteomics, and other experimental techniques, there has been an explosion of new protein structures being deposited in the databases. In many cases, however, these proteins have little or no functional annotation. Sequence-based approaches still remain the most effective way to assign function based on homology, but in cases of extreme divergence and analogous proteins these methods can fail. In order to identify these types of relationships, a number of structure-based approaches have been developed, such as the MSDmotif service. No single method is successful in all cases and a more prudent approach involves the utilization of data from a wide range of resources. One such approach is the ProFunc server, developed to help researchers narrow down the number of functional possibilities for experimental validation.

Keywords: structural genomics; function from structure; 3D motifs; bioinformatics.

165

R.J. Read and J.L. Sussman (eds.): Evolving Methods for Macromolecular Crystallography, 165–179
© 2007. *Springer*

1. Introduction

As the various global structural genomics projects have picked up pace the number of structures annotated in the Protein Data Bank [1] as "hypothetical protein" or "unknown function" has grown significantly. A major challenge in bioinformatics involves the development of computational methods for accurate and automatic assignment of functions to these proteins. The primary source for functional inference comes from sequence similarity, but when this process fails, analysis of the protein structure can provide functional clues. Many structure-based approaches exist from global fold similarity comparisons down to highly specific three-dimensional (3D) templates.

One such structural method developed at the European Bioinformatics Institute (EBI) is MSDmotif [2, 3]. This is a database of structural motifs including functional 3D motifs (active sites, catalytic sites, Prosite [4] patterns and profiles), small structural motifs (Beta turns, nests) and standard secondary structure elements like helix, strands, and turns. However, asides from being a useful database, MSDmotif has additional functionality that allows it to be used as a research tool to help identify and define new motifs that may have functional or structural significance.

This is but one of many methods for structural comparison and motif identification. Every method has its own success rate and some are family specific, so no single method can be expected to successfully predict a protein's function in all cases. A more prudent approach involves combining multiple methods and many such meta-servers are being developed. As part of the Midwest Center for Structural Genomics (MCSG) the EBI has developed a fully automated functional analysis server, ProFunc [5], which performs a battery of analyses on a submitted structure. The analyses combine a number of currently available sequence-based and structure-based methods to identify functional clues. These are presented to the user as a list of the top matches from each service thus leaving the final assessment of function open to the user with their expert knowledge of the query protein. There are substantial difficulties in any attempt to define a protein's function and, although there are a number of successful cases where structure-based predictions have been checked experimentally and a function identified, there are many cases where all analyses fail and the protein's function remains unknown.

2. Introduction to MSDmotif

2.1. OUTLINE OF METHOD

The aim of MSDmotif is to provide:

- Powerful 3D search tightly integrated with a sequence search
- Wide range of visualizations in 3D and two-dimensional (2D)
- Multiple sequence and structure alignments

The target of a search is a 3D motif. Loaded and mapped to the MSD database are 3D motifs, such as active sites, catalytic sites, Prosite patterns and profiles, and small motifs (derived by Professor James Milner-White) along with the standard secondary structure elements like helix, strands, and beta turns. MSDmotif is not only a valuable source of information, but also a research tool, which can help to identify new structural motifs.

MSDmotif is the first online service that provides a search by backbone parameters like phi (ϕ) and psi (ψ). These parameters with low variance can be translated into geometrical characteristics and vice versa. A ϕ/ψ search is linearly dependent on the number of residues and therefore has a huge advantage compared with the exponential dependency of a 3D search mechanism. Just to give a flavor of its advantages, a ϕ/ψ search by five residues in a sequence with tolerance of 30° on each ϕ, ψ takes only a matter of seconds.

The example illustrated below uses the calcium-binding loop from PDB entry 1gci (which belongs to the Subtilases family) and performs a search using the ϕ, ψ angles of its residues. The result gives not only other members of the Subtilases family, but also find hits from members of the Globins protein family. The 3D alignment of these loops (as a superposition of the two SCOP families) is presented in the square box of Figure 1.

2.2. SEARCHING WITH MSDmotif

There are a number of predefined small motifs in MSDmotif derived from data provided by Professor James Milner-White. The most famous example is the beta–turn, but other less-well-known motifs are also described. Although these predefined motifs exist, MSDmotif is not restricted to them and user-defined motifs can be searched for. The definition of such motifs

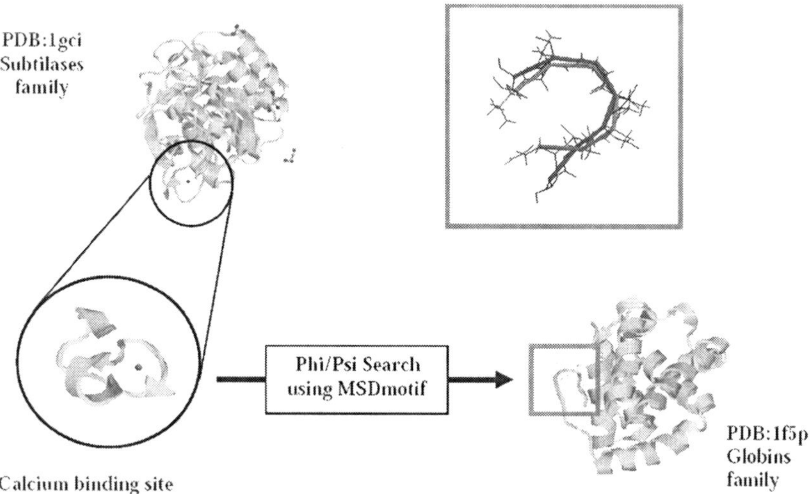

Figure 1. MSDmotif search using the calcium-binding loop of 1gci.

could consist of residue specifications, hydrogen bonds within main chain or side chain, or even restrictions on main chain dihedral angles (phi, psi, and chi). Any one or a combination of these parameters can be specified as search criteria for MSDmotif and in such a way new motifs could be derived from a structure or existing motif definitions can be expanded. As an example, the ST-staple definition states that first residue of this motif must be Serine or Threonine, but a search using similar ϕ/ψ angles shows that many other amino acids could be considered as a first residue of the motif.

For each small motif a number of statistics and charts are provided, including:

— Ramachandran plots for each residue of the motif

— 3D chart of amino acid occurrence at each position

— 3D chart of correlation between side chain charge and residue position

— 2D charts of motif parameter variation

— Ligand fragment binding statistics for each residue of the motif

Other options allow any matching fragments to a motif to be used to create a 3D alignment that is easily visualized. Figure 2 shows the alignment of the *ST-staple* motif found for different SCOP families and illustrates that although sequences can diverge the core structure can remain similar.

Another interesting search would be to find different protein families whose sequences comply with a given pattern. Here, the interest lies in finding distant relatives who have kept key residues hopefully essential to the function. MSDmotif implements an advanced database technique for this kind of search providing answers to most pattern queries within a couple of seconds. As an example, we can consider a Zinc finger C2H2 type domain signature pattern:

$$C - x(2,4) - C - x(3) - [LIVMFYWC] - x(8) - H - x(3,5) - H.$$

A more complex search would be to combine structural and sequence restraints, such an example would be to take a beta hairpin (strand–turn–strand) and impose the additional restriction that a glycosylation pattern is required within two or three residues after the motif, plus the filter that the first residue of the glycosylation pattern (Asparagine) must interact with a sugar residue (e.g., Mannose). Schematically this type of search criteria could be presented as in Figure 3.

A query using these criteria currently takes about 5 s. The results can then be superposed and aligned by secondary structure patterns. This would help researchers identify similarities that would be impossible to find using the sequence details alone. The use of sequence data and structural information can be essential to the elucidation of biochemical function.

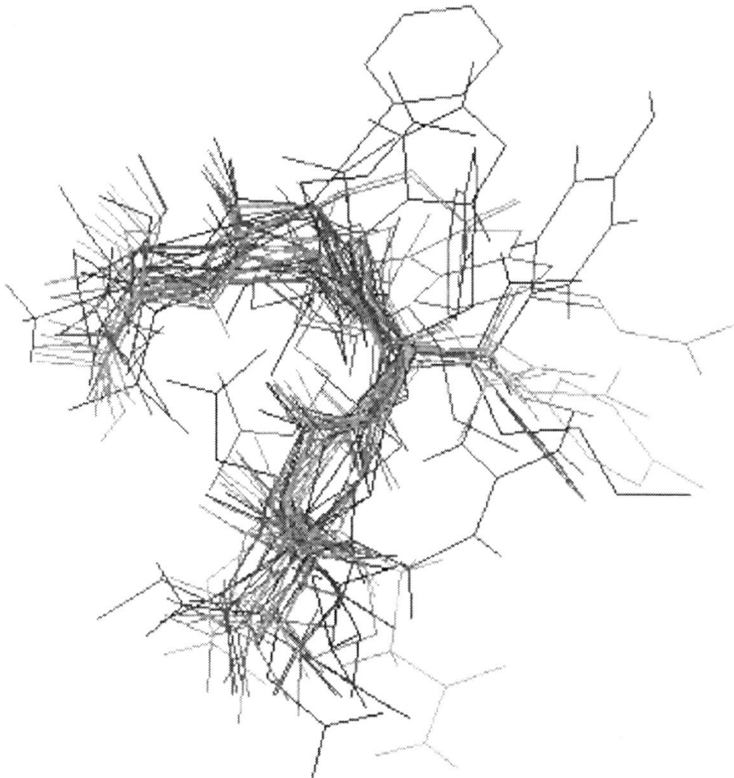

Figure 2. 3D alignment of ST-staple motifs from different SCOP families.

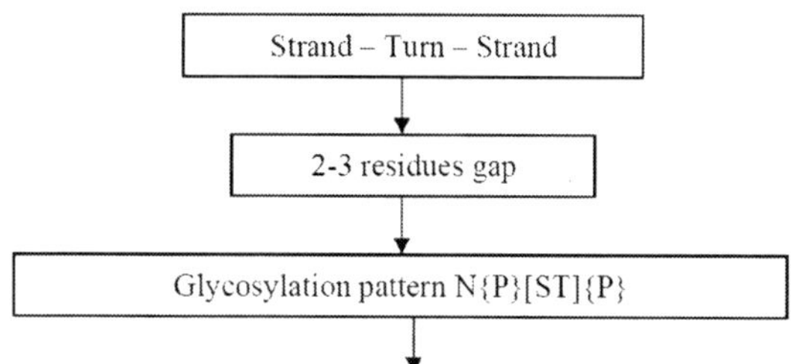

Figure 3. Example of a query for MSDmotif.

3. Introduction to ProFunc

3.1. OUTLINE OF METHOD

The most effective way of assigning a protein's function is by comparing the amino acid sequence with the sequence databases to find remote homologues of known function. Where suitable levels of sequence similarity exist (global similarity or local characteristic short patterns) it can be possible to group related proteins and infer similarity in function. The sequence methods have been improving over the years and modern statistical methods such as Hidden Markov models (HMMs) [6] can now successfully identify remote homologues below 30% global sequence identity. There are however, a significant number of proteins where the sequence methods provide no functional clues and in these cases we must look to the 3D structure to identify function. This involves numerous analyses since protein structure can be examined at different levels, ranging from the overall fold and protein–protein interaction level down to specific clusters and conformations of a handful of functional amino acids.

The ProFunc server (http://www.ebi.ac.uk/thornton-srv/databases/Pro Func/) was developed as part of a collaborative project with the MCSG and BioSapiens to produce a fully automated functional analysis pipeline. ProFunc performs a wide range of analyses (both sequence and structure based) on a submitted protein structure with an aim to identify functional clues. No single method will provide 100% accuracy, so the greater is the number and range of methods employed, the greater is the chance that at least one method will pick up a significant match to the "correct" function.

3.2. METHODS USED IN ProFunc

ProFunc methods can be split into two broad categories that currently consist of:

"Sequence-based" approaches:

- InterProScan
- FASTA vs PDB
- BLAST vs UniProt
- Residue conservation analysis and mapping onto structure
- Genome analysis
- Superfamily [10] search (SCOP HMM model library)

"Structure-based" approaches:

- Fold matching using SSM and DALI
- HTH motif scanning

- Surface cleft analysis
- Nest analysis
- 3D templates

3.2.1. Sequence-based methods

The use of BLAST [7] and FASTA [8] searches against the PDB and UniProt [9] is the fastest way to identify any homologous proteins. Generally speaking, any proteins with sequence similarity greater than 40–50% are likely to have similar structures and function, but there are extreme cases where this is not true [11] so care must be taken in interpreting the results. More confident functional associations are described by InterPro [12] (essentially a "database of databases" containing protein families, domains, and functional sites) in which identifiable features found in known proteins can be applied to protein sequences of unknown function. The most famous of the components are the Prosite and Pfam [13] databases.

One of the most important uses of sequence similarity is in the identification of the most important parts of a protein. Those residues performing catalytic reactions, binding of cofactors/metals, or those essential to creating protein–protein or protein–DNA interactions are often retained throughout evolution and are highly conserved residues. The calculation of conservation involves finding all sequences from the database similar to the one in question, create a multiple sequence alignment to get the best possible alignment (using CLUSTALW [14]), calculating the variability across these sequences, and finally calculating a conservation score [15]. This can then be mapped onto the protein and used to identify conservation "hot spots" on the protein (potential active sites or interaction surfaces [16]).

In many organisms (especially bacteria), genes that act within the same biochemical pathway or act under the same environmental conditions are located in the same section of the genome (e.g., the lactose operon of *Escherichia coli*). The ProFunc server identifies the genomic location of any highly sequence homologues, constructs a diagram of the genome organization and displays any functional information about the surrounding genes. If the gene of interest is part of an operon, the surrounding genes should have functions relating to the overall pathway. Another use of this information is to compare the organization of genes surrounding that of interest *across* all genomes in which a homologue is found. In this case, it then becomes possible to identify co-occurring genes that may or may not be functionally related.

3.2.2. Structure-based methods

There are various "levels" on which to consider protein structure. At the large-scale end it is important to consider how the multitude of proteins interact with one another in the cell to perform their functions (e.g., looking at protein–protein interaction networks, regulatory networks, metabolomics, and gene expression analysis). The next step down is to consider the multimeric state of the submitted structure. Multimeric complexes can be homomeric (all subunits the same) or heteromeric (subunits differ from one another) in nature and the number of subunits can range from 2 to 60 (and possibly even more). ProFunc therefore offers an assessment of the most likely biologically significant multimeric state (using the Pita [17] software) and subsequent analyses can then be performed on the multimer or the single chain. The structural analyses performed by ProFunc are listed in order of increasing focus and local level of comparison:

- Fold recognition
- Supersecondary motifs
- Surface cleft analysis
- Small structural motifs
- 3D templates

3.2.3. Fold recognition

It has been estimated that although the number of possible protein sequences is almost limitless the number of folds that they make up is much more restricted. By comparing folds distant evolutionary relationships can be identified and cases of convergent evolution found (where a functional fold is evolved independently from different ancestral proteins). There are a number of tools available to compare proteins and identify fold similarity the most famous being the DALI [18] server, CE [19] (combinatorial extension), and VAST [20] (Vector Alignment Search Tool). ProFunc uses the SSM [21] (secondary structure matching) algorithm developed at the EBI in addition to a DALI run. In general, both methods provide similar results but because neither method is 100% accurate, the inclusion of both lowers the risk of missing similarities and allows the user to decide which is the best match.

3.2.4. Supersecondary motifs

The next stage down from the overall fold is to look at subsections of fold for motifs with functional significance. These can be considered supersecondary motifs because they usually consist of two or more parts of secondary structure (e.g., two helices, helical bundles, two strands forming a hairpin). The most significant example of these motifs is that of the helix–turn–helix (HTH) DNA-binding motif. All of these motifs can also be identified by the MSDmotif service.

Examination of DNA-binding proteins identified a number of super-secondary motifs used by the proteins to interact with the DNA:

- Helix–turn–helix (HTH)
- Helix–loop–helix (HLH)
- Helix–hairpin–helix (HHH)
- Winged-helix (WH)

The HTH motifs were shown to be the most amenable to structural characterization and therefore more likely to be predictable. Structural templates are used to assess new structures for the presence of the HTH motif. In order to identify only those that are true DNA binders, the solvent accessibility and electrostatic potential of the match is also calculated: true DNA-binding HTH motifs should be solvent accessible and have a largely positive electrostatic potential. The use of these two additional parameters greatly increases the predictive power of the method [22].

3.2.5. Surface cleft analysis

Studies into enzymes have identified that the active site is most commonly located in one of the two largest surface clefts. Natural substrates, pharmaceuticals, and regulatory molecules bind to pockets on the surface either blocking action of a protein or altering the mode of action. Identifying the surface clefts and binding pockets is therefore essential to the assessment of potential functions.

ProFunc uses the SURFNET [23] algorithm, to calculate the ten largest clefts for display to the user. The surfaces of each pocket can be colored by the residue conservation score or by the physicochemical properties of the atoms contributing to the pocket. Large pockets with highly conserved residues are more likely to be functionally important and can be used to help identify functional residues. Looking at the properties of the residues lining the pocket can also help narrow down the list of potential substrates or cofactors.

3.2.6. Small structural motifs

To look in more detail at the structure involves moving down to the short loops and small hydrogen-bonded motifs that make up the secondary structure of the protein. One such motif is termed the "nest" [24, 25] (one of the predefined entries in MSDmotif database). Where regular secondary structures are formed by successive amino acids having identical conformations, nests occur where successive amino acids have alternating right- and left-handed conformations. The resultant conformation forms a slightly concave depression or "nest" of backbone amide groups. The concavity has a slightly positive nature and therefore is capable of binding small anions, main chain carbonyl groups or amino acid side chains. Multiple nests can arrange themselves into much larger repeating structures capable of binding larger anionic compounds such as ATP. Nests play structural roles (such as terminating alpha helices and forming beta turns) and are found in a number of functional sites (such as the P-loop of ATP-binding proteins, Iron–sulphur binding sites, and the oxyanion holes of serine proteases). ProFunc identifies any nests in the structure and ranks them by the length of the nest, the average conservation of residues involved and location in or near any of the largest clefts.

3.2.7. Three-dimensional templates

Much of the functional chemistry performed by enzymes and recognition of substrates is achieved a few highly conserved residues. Sometimes the residues are conserved not in sequence but in 3D location and although the rest of the protein structural scaffold changes, the essential catalytic residues are retained in the correct orientation. The most famous and extensively characterized example of this is that of the Ser–His–Asp catalytic triad. The catalytic mechanism of the Ser–His–Asp triad is used in a number of different folds, but the chemistry of their components remains the same. This activity depends on the 3D arrangement of the three residues and any deviation from this disrupts the function.

Other enzymes use alternative arrangements of other residues to perform a wide variety of catalytic reactions. A database of catalytic sites, the Catalytic Site Atlas (CSA) [26], was developed over a number of years to develop 3D templates for each type of catalytic mechanism described in the literature. There are now over 200 different templates in the database covering a wide range of enzyme reactions.

A rapid template-scanning algorithm, JESS [27], was developed to rapidly scan a submitted structure for the presence of any given template and calculate the root-mean-square deviation (rmsd) of the match. The use of rmsd on its own often returns a large number of false positive matches, so in order to increase the specificity a 10 Å sphere is drawn around the hit and the template structure. If a hit is a true hit, it is expected that the local environment around the template match should be the same for both proteins (similar active sites). By comparing the surrounding residues for chemical similarity and the distribution of empty space between the hit and the original, it is possible to filter out the obvious false positives from the true hits of interest.

There are four types of template scans performed:

- Enzyme templates
- Ligand-binding templates
- DNA-binding templates
- "Reverse" templates

Enzyme templates from the CSA are manually generated and topped up by semiautomatically generated templates recently created to expand the coverage of the data set [28]. These are high-confidence and high-specificity templates as many have been manually derived from extensive literature searches and analysis of catalytic mechanisms.

Ligand-binding templates are automatically generated from the structures deposited in the Protein Data Bank. For every ligand entry in the PDB a nonhomologous data set of proteins containing that ligand is constructed. Residues interacting directly with the ligand are used to construct a number of three-residue templates where residues are within a certain distance cutoff from one another and no two templates have more than one residue in common. The DNA-binding templates are constructed in the same manner, but utilize a data set of nonhomologous DNA-binding proteins from the PDB.

The "reverse template" [29] approach turns the idea of templates on its head. Rather than scanning a single query protein for hits to a database of templates of known function, this method generates templates from the query protein and scans each template in turn against a nonredundant representative sample of the entire PDB. This method is more successful at identifying relationships than other methods because the template-generation process makes few assumptions. The main criteria for selection of residues is that they should be highly conserved, within a distance cutoff from one another and they do not form part of the central core of the protein (matching cores of proteins is uninformative and can be best achieved using fold recognition algorithms). On average a submitted protein structure will generate over 100 different templates each of which needs to be scanned for matches to any of the thousands of structures in the PDB. In order to achieve this computationally intensive task a processor farm runs multiple jobs in parallel increasing the efficiency significantly. The local environments of hits are then checked and a statistical scoring scheme is employed to rank the matches by significance. Each hit can be viewed in turn and a superposition of the structures based on the local environment match is made available (e.g., that shown in Figure 4).

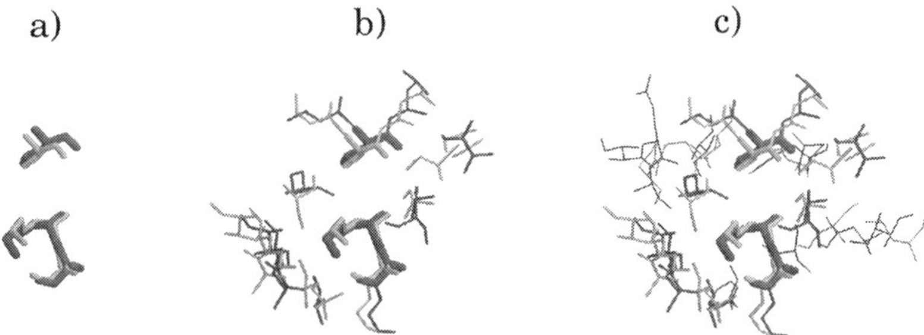

Figure 4. The template local comparison. (a) The superposed template residues, the dark residues being from the template and the light residues from the query structure. (b) As in (a), but with residues of identical residue type, lying within 10 Å of the template center, that overlap when the template and query structures are superposed. (c) As in (b), but with overlapping residues of similar residue type added. The more residues one gets in (b) and (c), the more similar are the local environments of the template match and hence the stronger the certainty of the match.

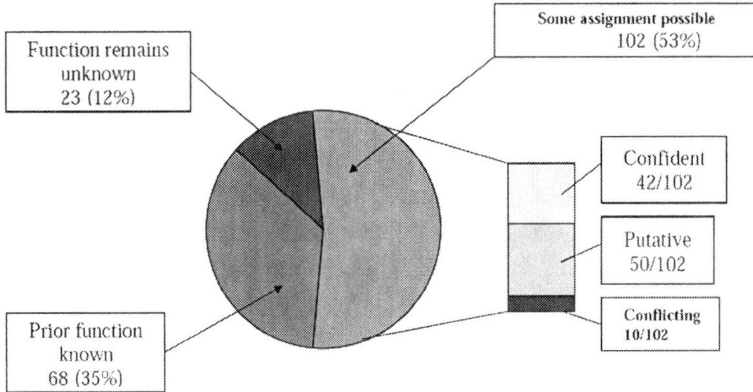

Figure 5. Summary of functional annotation of MCSG structures (193 structures).

3.3. EXAMPLES USING ProFunc

The MCSG has been essential to the development of ProFunc and has supplied an enormous number of structures to test the methods on (258 structures as of 30 March 2005). A pipeline is now in place to automatically generate ProFunc output and PDBsum pages for each structure as it is solved. There are a number of successful cases where structure-based predictions have been checked experimentally and a function identified, but there are also cases where all of the analyses fail. A summary of the predictive ability of ProFunc was assessed using 193 of the MCSG structures and the results are illustrated in Figure 5.

3.3.1. Good example (BioH)

The BioH protein from *E. coli* (PDBcode: 1m33) was annotated as a hypothetical protein believed to be involved in Biotin biosynthesis. Analysis of the sequence provided few functional clues so it was hoped that analysis of the structure would provide clues to its function. A ProFunc run identified a surprisingly strong enzyme template match (rmsd of only 0.28 Å) to a Ser–His–Asp catalytic triad, suggesting a hydrolase function. This functional prediction was experimentally confirmed and BioH is now known to be a novel carboxylesterase acting on short acyl chain substrates [30].

3.3.2. Failed Example (Function Remains Unknown)

YjcS from *Bacillus subtilis* (PDBcode: 1Q8B) is annotated as a structural genomics protein of unknown function. All sequence-based approaches in ProFunc provide little or no information: there are no sequence motifs from the InterProScan, the BLAST runs hit only hypothetical proteins and there is poor residue conservation across the sequence. The structure-based approaches provide as little information as the sequence-based ones. The fold comparison servers suggest it is most similar to hypothetical proteins and plant "stable proteins." There are no significant template hits to enzymes, ligands or DNA. There are however, a number of significant reverse template hits, but all of them are to other structural genomics proteins of unknown function. In this case, further experiments are required to determine the function.

4. Conclusions

The development of computational methods for accurate and automatic assignment of functions to proteins is key to the success of structural genomics initiatives. Improvements to sequence similarity methods allow for comparisons of distantly related proteins and new structure-based methods can provide functional clues unavailable from the sequence alone. One such method, MSDmotif, is the first online service that provides a search by backbone parameters and allows for rapid powerful searches of proteins for well-known functional motifs. Additional searches allow researchers to identify and define new motifs with functional and structural significance.

Small 3D motifs are only one aspect of protein structure and in order to accurately predict a protein's function one must take into account information from a wide variety of sources such genome organization, fold class, protein–protein interaction partners, protein–DNA interactions, subcellular localization, expression profiles, and others. Methods utilizing information from multiple sources are under active development by many groups. One such server, ProFunc, performs a combination of sequence-based and structure-based methods to identify functional clues. Not every method works in every case, and examples abound where structure-based function predictions have been confirmed as well as those where all analyses fail. An analysis of 193 structures solved by the MCSG suggests that, despite all efforts, over 12% of structures remain classed as "function unknown".

References

1. Berman, H.M., Westbrook, J., Feng, Z., Gilliland, G., Bhat, T.N., Weissig, H., Shindyalov I.N., and Bourne, P.E. (2000) The Protein Data Bank. *Nucleic Acids Research*, **28**: 235–242.
2. Golovin, A., Dimitropoulos, D., Oldfield, T., Rachedi A., and Henrick, K. (2005) MSDsite: A Database Search and Retrieval System for the Analysis and Viewing of Bound Ligands and Active Sites. *Proteins: Structure, Function, and Bioinformatics*, **58**(1): 190–199.
3. Golovin, A. (2004) MSDmotif: a database search and retrieval system for the analysis and viewing of protein structure motifs. The eCheminfo 2005 Conference "Webservices" 13 June.
4. Hulo, N., Sigrist, C.J.A., Le Saux, V., Langendijk-Genevaux, P.S., Bordoli, L., Gattiker, A., De Castro, E., Bucher, P., and Bairoch, A. (2004) Recent improvements to the PROSITE database. *Nucleic Acids Research*, **32**: D134–D137.
5. Laskowski, R.A., Watson, J.D., and Thornton, J.M. (2005) ProFunc: a server for predicting protein function from 3D structure. *Nucleic Acids Research*, **33**: W89–W93.
6. Karplus, K., Barrett, C., and Hughey, R. (1998) Hidden Markov models for detecting remote protein homologies. *Bioinformatics*, **14**: 846–856.
7. Altschul, S.F., Madden, T.L., Schaffer, A.A. Zhang, J., Zhang, Z., Miller, W., and Lipman, D.J. (1997) Gapped BLAST and PSI-BLAST: a new generation of protein database search programs. *Nucleic Acids Research*, **25**: 3389–3402.
8. Pearson, W.R. (1998) Empirical statistical estimates for sequence similarity scores. *Journal of molecular biology*, **276**: 71–84.
9. Apweiler, R., Bairoch, A., Wu, C.H., Barker, W.C., Boeckmann, B., Ferro, S., Gasteiger, E., Huang, H., Lopez, R., Magrane, M. Martin, M.J., Natale, D.A., O'Donovan, C., Redaschi, N., and Yeh, L.S. (2004) UniProt: the Universal Protein Knowledgebase. *Nucleic Acids Research*, **32**: D115–D119.
10. Madera, M., Vogel, C., Kummerfeld, S.K., Chothia, C., and Gough, J. (2004) The SUPERFAMILY database in 2004: additions and improvements. *Nucleic Acids Research*, **32**: D235–D239.
11. Whisstock, J.C., and Lesk, A.M. (2003) Prediction of protein function from protein sequence and structure. *Quarterly Reviews of Biophysics*, **36**(3): 307–340.
12. Mulder, N.J., Apweiler, R., Attwood, T.K., Bairoch, A., Barrell, D., Bateman, A., Binns, D., Biswas, M., Bradley, P., Bork, P., et al. (2003) The InterPro Database, 2003 brings increased coverage and new features. *Nucleic Acids Research*, **31**: 315–318.
13. Bateman, A., Birney, E., Cerruti, L., Durbin, R., Etwiller, L., Eddy, S.R., Griffiths-Jones, S., Howe, K.L., Marshall, M., and Sonnhammer, E.L. (2002) The Pfam protein families database. *Nucleic Acids Research*, **30**(1): 276–280.
14. Higgins, D., Thompson, J., Gibson, T., Thompson, J.D., Higgins, D.G., and Gibson, T.J. (1994) CLUSTAL W: improving the sensitivity of progressive multiple sequence alignment through sequence weighting, position-specific gap penalties and weight matrix choice. *Nucleic Acids Research*, **22**: 4673–4680.
15. Valdar, W.S.J., and Thornton, J.M. (2001) Conservation helps to identify biologically relevant crystal contacts. *Journal of Molecular Biology*, **313**: 399–416.

16. George, R.A., Spriggs, R.G., Bartlett, G.J., Gutteridge, A., MacArthur, M.W., Porter, C.T., Al-Lazikani, B., Thornton, J.M., and Swindells, M.B. (2005) Effective function annotation through residue conservation. *Proceedings of the National Academy of Sciences of the USA*, **102**: 12299–12304.

17. Ponstingl, H., Kabir, T., and Thornton, J. M. (2002) Automatic inference of protein quaternary structure from crystals. *Journal of Applied Crystallography*, **36**: 1116–1122.

18. Holm, L., and Sander, C. (1995) Dali: a network tool for protein structure comparison. *Trends in Biochemical Sciences*, **20**: 478–480.

19. Shindyalov, I.N., and Bourne, P.E. (1998) Protein structure alignment by incremental combinatorial extension (CE) of the optimal path. *Protein Engineering*, **11**: 739–747.

20. Madej, T., Gibrat, J.F., and Bryant, S.H. (1995) Threading a database of protein cores. *Proteins*, **23**(3): 356–369.

21. Krissinel, E., and Henrick, K. (2004) Secondary-structure matching (SSM), a new tool for fast protein structure alignment in three dimensions. *Acta Crystallographica*, **D60**: 2256–2268.

22. Ferrer-Costa, C., Shanahan, H.P., Jones, S., and Thornton, J.M. (2005) HTHquery: a method for detecting DNA-binding proteins with a helix-turn-helix structural motif. *Bioinformatics*, **21**: 3679–3680.

23. Laskowski, R.A. (1995) SURFNET: a program for visualizing molecular surfaces, cavities and intermolecular interactions. *Journal of Molecular Graphics*, **13**: 323–330.

24. Watson, J.D., and Milner-White, E.J. (2002) A novel main-chain anion-binding site in proteins: the nest. A particular combination of phi, psi values in successive residues gives rise to anion-binding sites that occur commonly and are found often at functionally important regions. *Journal of Molecular Biology*, **315**(2): 171–182.

25. Watson, J.D., and Milner-White, E.J. (2002) The conformations of polypeptide chains where the main-chain parts of successive residues are enantiomeric. Their occurrence in cation and anion-binding regions of proteins. *Journal of Molecular Biology*, **315**(2): 183–191.

26. Porter, C.T., Bartlett, G.J., and Thornton, J.M. (2004) The Catalytic Site Atlas: a resource of catalytic sites and residues identified in enzymes using structural data. *Nucleic Acids Research*, **32**: D129–D133.

27. Barker, J.A., and Thornton, J.M. (2003) An algorithm for constraint-based structural template matching: application to 3D templates with statistical analysis. *Bioinformatics*, **19**: 1644–1649.

28. Torrance, J.W., Bartlett, G.J., Porter, C. T., and Thornton, J.M. (2005) Using a library of structural templates to recognise catalytic sites and explore their evolution in homologous families. *Journal of Molecular Biology*, **347**: 565–581.

29. Laskowski, R.A., Watson, J.D., and Thornton, J.M. (2005) Protein function prediction using local 3D templates. *Journal of Molecular Biology*, **351**: 614–626.

30. Sanishvili, R., Yakunin, A.F., Laskowski, R.A., Skarina, T., Evdokimova, E., Doherty-Kirby, A., Lajoie, G.A., Thornton, J.M., Arrowsmith, C.H., Savchenko, A., Joachimiak, A., and Edwards, A.M. (2003) Integrating structure, bioinformatics, and enzymology to discover function – BioH, a new carboxylesterase from *Escherichia coli. The Journal of Biological Chemistry*, **278**: 26039–26045.

SINGLE-PARTICLE IMAGING

DAVID SAYRE

*Department of Physics and Astronomy, Stony Brook
University, Stony Brook, NY 11794, USA*

Abstract: The current status of single-particle X-ray imaging – or "X-ray crystallography without need for a crystal" – is reviewed, with particular emphasis on the Stony Brook/Cornell/Berkeley single yeast cell imaging project. Two-dimensional (2D) images of a quick-frozen yeast cell have been obtained with a resolution which, if carried out in three-dimensional (3D), would allow observation of the positions of the large molecular assemblies in an entire almost-living-state cell. The projected shift from 2D to 3D imaging is discussed.

Keywords: X-ray single-particle imaging; X-ray diffraction microscopy; yeast cell imaging; X-ray flash imaging of protein molecules.

1. Introduction

The subject of this paper is the imaging of general small objects by essentially the same methods as are used today for the imaging of unit cells of crystals; it could thus be called "crystallography without need for a crystal". Here is the point we had reached in the subject 5 years ago, when I gave a talk on it in Erice in 2000, describing work published in the previous year [1]. The "small object" looked like Figure 1a under a scanning electron microscope (SEM). Its diffraction pattern when placed in the X-ray beam looked like Figure 1b, and the image produced (i.e., the Fourier transform of the pattern after phasing) looked like Figure 1c. Note that the diffraction pattern was not concentrated into discrete Bragg spots, as it would have been if the object had been repeated many times on a lattice. It was much weaker than that, and required more intense exposure to be recorded, but being continuous (another picture below will let you see the continuity much better), and not having lost the information *between* the Bragg spots, it carried more information – enough in fact that it only took a rough knowledge of the *envelope* of the object to allow the phases to be quite easily determined. I will go into the method of phasing in more detail later.

R.J. Read and J.L. Sussman (eds.): Evolving Methods for Macromolecular Crystallography, 181–191
© 2007. *Springer*

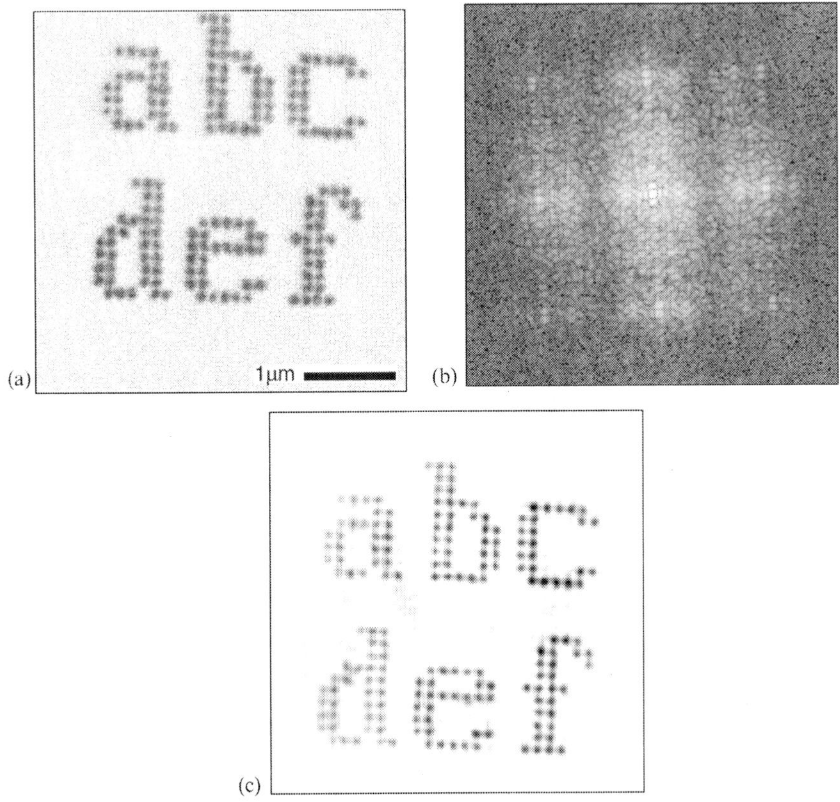

Figure 1. Status of the field 5 years ago [1].

Today, we have advanced much further, to the imaging of a whole biological cell, a yeast cell, recently living, but then plunged into liquid ethane, and then brought up through temperature stages to sublimate out its water content and become at room temperature a freeze-dried cell [2]. Many of its fellow cells, when put back in water, resume living. Its diffraction pattern, taken with the specimen sitting stationary in a 16.5 Å wavelength X-ray beam at the Lawrence Berkeley ALS by (at that time) Stony Brook graduate student David Shapiro is shown in Figure 2. You can see the continuity of the pattern. The transform of the pattern, with the phasing mainly worked out by Professor Veit Elser at Cornell and his graduate student Pierre Thibault, is shown in Figure 3. The colorization here is genuine, in the sense that hue and intensity code for the real and imaginary parts of the reconstructed density. Let me point out the scale bar, which tells us that the cell is about 3 μm in diameter, and that repeatable detail in it, down to the 30 nm size level, is present.

Thus, since 30 nm is about the size of an individual ribosome, this picture suggests that the day is not far off when a three-dimensional (3D) image like this will allow us to see the position of every object of ribosome or larger size in a cell, and to do so in a cell that was living up to the moment of its freezing. David's data actually go out to a little better than 10 nm resolution. Thus, developing the imaging to three dimensions and to that higher resolution is the next major goal of our group, and I will be discussing that later. What might be the impact of such imaging in biology? Just as the knowledge, provided by crystallography, of atomic positions and movements has allowed a detailed understanding of the actions of large biomolecular entities, knowledge of large biomolecular entity positions and movements could lead to a similar understanding of the actions of entire cells.

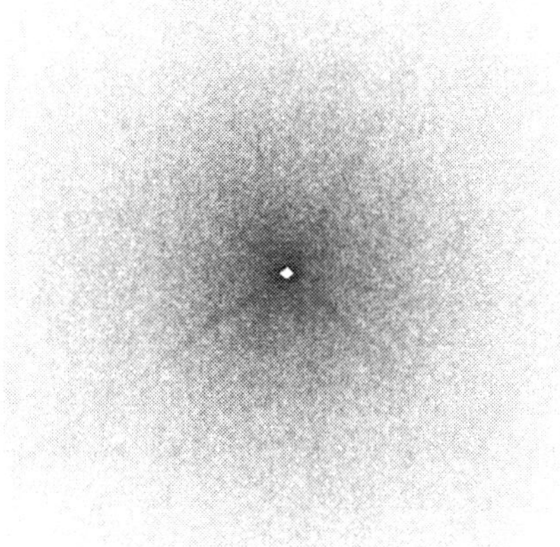

Figure 2. Current status of the field [2]. Two-dimensional (2D) diffraction pattern of a quick-frozen freeze-dried yeast cell.

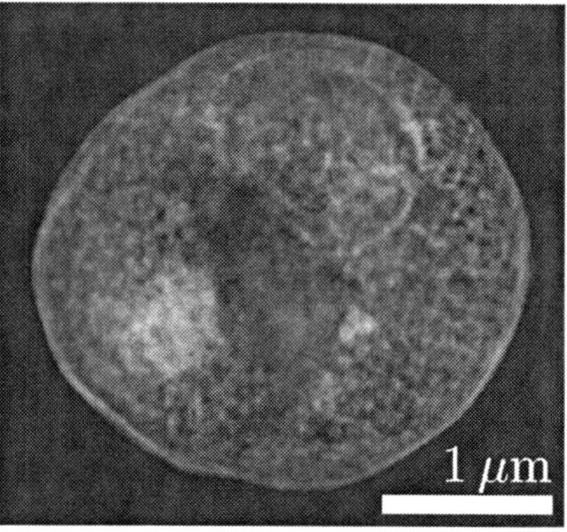

Figure 3. Current status of the field [2]. Two-dimensional (2D) image of the cell.

This work on the cell is the work which our group is doing, but a lot more is now going on. The methodology today is starting to be known as X-ray single-particle imaging or X-ray diffraction microscopy, and people now are working on it in more than a dozen institutions around the world, including Stony Brook and Cornell in the USA east coast, Illinois in the Midwest, and Stanford, Berkeley, Arizona State, Livermore, and UCLA in the west, as well as Uppsala in Sweden, Hamburg, and Berlin in Germany, and SPring8 in Japan. In some cases, the emphasis is on materials science specimens; in other places, work has turned to the possibility of imaging biological macro-molecules and macromolecular assemblies, i.e., as here at this school, but via single particles instead of crystals. At the same Erice meeting 5 years ago, Janos Hajdu gave his paper on the possibility of diffracting from individual protein molecules placed in the beam path of a femtosecond pulse-length X-ray source, and the coming together of our paper and his paper led to discussions right here at Erice at which the projects for single-particle bio-molecular imaging which now exist at Stanford and at HASYlab received I think their launching. Five years on, single-particle prospects now exist in biomolecular, biocellular, and materials science areas.

2.   Carrying out single-particle imaging

Since everyone here is familiar with how crystallography is done, the simplest thing will be to bring out the differences which exist, so I will do that, under the three main headings of getting the specimen, getting the intensities, and getting the phases.

2.1. GETTING THE SPECIMEN AND MOUNTING IT FOR EXPOSURE

Here is where the most important difference comes, in that getting a good crystal is not required. Grab a cell, or a molecule, or any tiny fragment of matter, and it will be a fine specimen. This can bring a very large saving of time and at the same time an opening up of new scientific application areas. It can be a problem too. For example, in our yeast cell work, because of the weakness of the scattering from the very small noncrystalline specimen, it is important that the amount of specimen mounting material in the beam be held to a minimum – a few microns of carbon nanofiber would be good, for example, or a microcapillary with nanothin carbon walls. At present, how-ever, we are still tied to the conventional planar geometry of a nanothin membrane supported on an EM grid, with the unwanted effect that when the grid is edge-on to the beam it blocks transmission and prevents access to some of the 3D data. (Fortunately, the phasing technique can go some distance in supplying that missing data.) In the molecular case, molecular spraying techniques promise to give a fully mountless method of getting a single molecule briefly into position for the femtosecond beam flash. That is very good, but of course raises the familiar problem of knowing, for each pattern obtained, what the molecular orientation was for that pattern; however, that problem is solvable provided the photons/pulse are sufficient to keep the Poisson noise level reasonable in the individual patterns. There may also be potential particle orientation techniques, using, e.g., polarized laser fields, for this problem. Summarizing, in obtaining specimens, single-particle work gives an extremely large increase in ease and range. But it gives a few new specimen-handling problems as well.

2.2. EXPOSURE AND RADIATION DAMAGE

Crystallography is very strong here, due to the reduction in X-ray exposure given by the large signal amplification at the Bragg spots, and due also to the sharing of damage over the many copies of the object being imaged. The amplification is lost in single-particle work. But where the many copies exist, single-particle work, operating through the concept of the sufficiently brief X-ray flash, should be able (when it is perfected) to image at atomic resolution, and indeed to do so whether or not the growth of a suitable crystal can be carried out. So in the biomolecular and biomolecular assembly situation, it may be that nothing at all will be lost by single-particle work, and indeed, in those cases today, where it is difficult or impossible to persuade the copies to assemble into a crystal, a great deal will actually be gained. Then too, in the opposite *no*-exact-copy situation – as today in the biological cell case and in many materials science cases – where by definition a crystal cannot exist for crystallographic study, single-particle imaging again will at least image. Then, in materials work, where specimens are often quite radiation resistant, and a single copy may survive in the X-rays long enough to give all the 3D data needed, single-particle imaging at atomic resolution may frequently be possible. Finally, however, when we come to the biocellular case, unless a population of cells identical all the way to the atomic level someday becomes available, imaging resolution must depend upon how long a single copy can continue to diffract consistently in the X-ray beam. Fortunately, resistance of a cell to damage can be increased through a number of techniques, of which the least invasive is instant fixation by fast freezing, followed if desired by the addition of chemical fixatives as well as of specific site markers, etc. Perhaps the most fundamental and satisfying single finding we have made thus far in our work on the yeast cell is that fast freezing, either by itself or followed by freeze-drying, gives enough radiation resistance to allow the collection of hundreds of good 10 nm resolution diffraction patterns from one cell; this means that for the yeast cell at least there should be no fundamental barrier to that level of 3D imaging.

2.3. PHASING AND IMAGE RECONSTRUCTION

Here too is an area where single particles do very well: in the crystal case, the transform of the unit cell is effectively observable only at the Bragg points, whereas in the single-particle case the transform, though weak, is continuous and can be observed at any desired fineness. Given the finer sampling, provided that a fairly good envelope of the particle is known, the phase problem almost disappears. It takes quite a lot of computation to make it disappear, but it does effectively disappear. This story developed over a number of years. It began with a paper, based on Shannon's theorem, that I wrote in 1952 on what would happen if we had finer-than-crystallographic sampling, followed by the work of Gerchberg and Saxton in the 1970s based on EM data, which does allow fine sampling; then the work of Bates and especially Fienup in the 1980s; then again by us and John Miao in the 1990s; and now finally by Elser, and Elser and Thibault, at Cornell. There are several different ways of presenting the subject. Pierre Thibault, in his poster at Erice, uses the language of iterated projections, but I will talk about it in ordinary crystallographic language. To arrive at the basic idea, imagine that we have sampled the magnitudes at more than Bragg fineness, and imagine also that we somehow have the correct phases. Fourier summing will then deliver the correct image repeated on a lattice larger than that which is necessary to keep the repeated images separate from each other, i.e., each image will be surrounded by a sea of zeroes. Given incorrect phases, however, some density will escape into what should be the sea of zeroes. Starting with random phases, then, go into image space, push all nonzero pixels outside the specimen envelope towards zero, come back into diffraction space, adopt the new phases, and repeat, until finally everything outside the envelope is zero; what is inside the envelope will be the correct structure. Here (Figure 4) is an early demonstration of the process, which we published in Acta in 1998 [3]. A finely sampled data set (Figure 4a) was given random starting phases and transformed (Figure 4b). The transform after 50 cycles of pushing down on the values outside the envelope is shown in Figure 4c. There is still a sprinkling of values which have not been fully pushed down. After a further 50 cycles (Figure 4d), the outside values now all look like zero — but things are still evolving. Figure 4e is the transform 100 cycles later — motion has now ceased, and the structure inside the envelope has reached its final and more meaningful form. Figure 4f shows the structure which was used in the first place to generate the data set. You cannot distinguish it from the final transform.

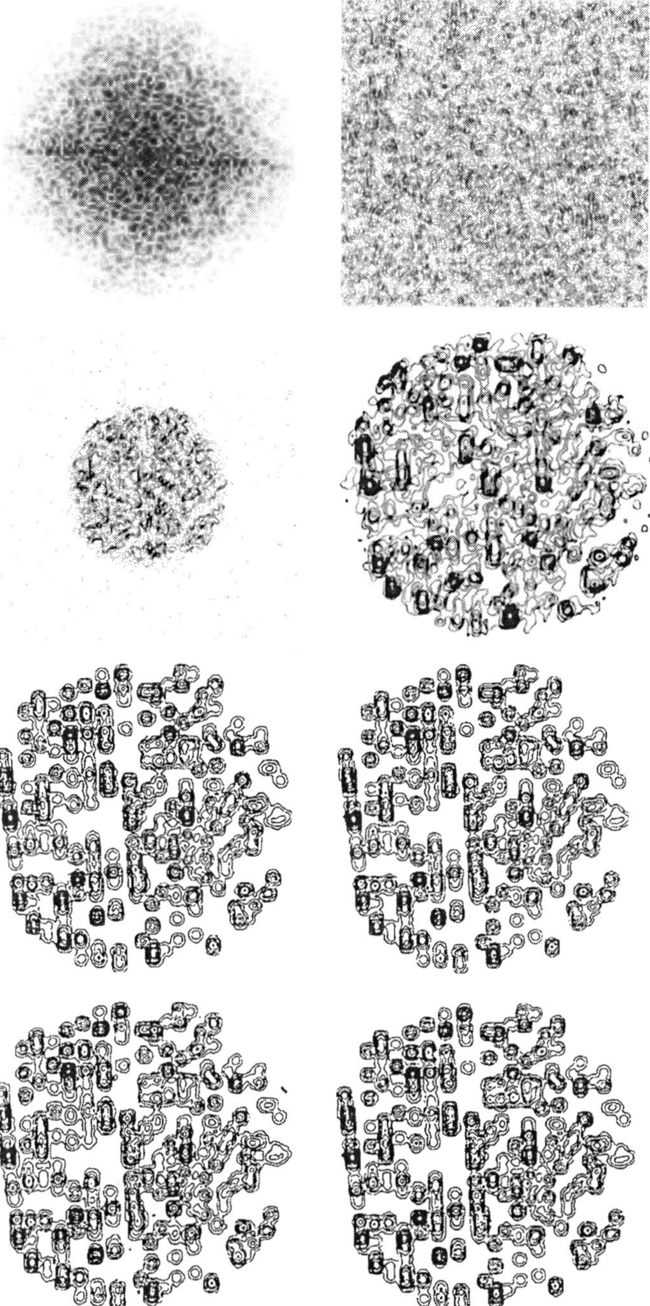

Figure 4. The principle of the phasing [3].

The last two figures (Figure 4g and h) show what happened when 17% peak-to-peak noise was added to corrupt somewhat the initial data set. It now took 425 cycles to reach a final structure, and it is now slightly visibly different from the transform of the changed magnitudes with the correct phases. The process is therefore fairly robust in the presence of experimental error. Detailed issues of how large the zero sea should be, of the best ways to push towards zero, of enforcing possible constraints on the pixels inside the envelope, and of refining the envelope itself, have been worked on by Fienup and others, and especially recently in the "difference map" technique of Elser. Today, the overall result is clear: given a reasonable initial knowledge of the envelope, reasonable sampling fineness, few sampling absences, reasonably small error in measuring the magnitudes, and plenty of computing power, the method is proving to be reliable, fast, and easy.

Taken all together, then, single-particle X-ray imaging depends on the existence today of more powerful X-ray beam technologies than heretofore and on methods for circumventing the damaging effects of increased X-ray exposure. In return, we obtain access to major new classes of specimens, and a very effective new approach to the phasing problem.

3. The yeast cell project

Now I would like to turn back to the specific project that I know best, the Stony Brook/Cornell/ALS Berkeley yeast cell project. This had its origins in the 1970s, and arose from the uncomfortable feeling that I had that crystallography must in time run out of crystals; I asked myself if there might be some way of letting X-rays image without crystals. Using ultrasmall Fresnel zone plates as X-ray lenses, Janos Kirz at StonyBrook and I began by working on the idea of X-ray microscopes, but in 1979, the thought came to me that perhaps the key ideas of crystallography itself could be made to work without crystals [4]. Brookhaven was building a synchrotron, and Janos was soon building a beamline and scanning X-ray microscope there, and through the 1980s and 1990s, he sheltered and encouraged the project, giving me photons and help of every kind, including almost always a graduate student or a postdoc. These years saw the first two crucial questions – could a synchrotron get a detectable diffraction pattern out of a single biological cell? Could that pattern be phased? – answered in the affirmative, the first by graduate student WenBing Yun in 1987 and the second starting in 1989 when Gerard Bricogne and I 1 day fished back in our memories to the paper I had written in 1952 [5] saying that if crystals would only let us measure between the Bragg spots, phasing would become much easier, and we realized that the noncrystal, which was what we wanted to work with anyway, was also the very thing that would indeed get us between the Bragg spots. I put that in a talk given here at Erice in 1990 [6], and following that, in 1995, postdoc Henry Chapman, who knew about Fienup's iterative work, and, in 1998, graduate student John Miao, then did actually demonstrate the hoped-for phasing [3]. Then finally in 1999, John Miao put the data-taking and the phasing together, and we did the 5-year-ago experiment that I showed at the start of this talk [1]. I turned 75 at about that time, and Janos Kirz and younger faculty member Chris Jacobsen at Stony Brook took over the management of the project, and were successful in obtaining a research grant from the National Institutes of Health (NIH) to try the method on an yeast cell. The crucial question now was whether the yeast cell could survive in the beam long enough to allow the measuring of a full 3D diffraction pattern. For this a new low-temperature apparatus needed to be built; the work was moved out to Berkeley for more photon intensity; and last year David Shapiro, helped by graduate students Enju Lima and Huijie Miao, gave an affirmative answer to the survival-length question, and also embarked on the transition to 3D imaging with the small eight-pattern rotation set of patterns taken at 1° rotation intervals which I have already mentioned. In the meantime, in 2002, I had become aware of the great strengthening that Veit Elser at Cornell was contributing to phasing technique, and in 2003–2004 we started to work in earnest with him and his student Pierre Thibault, who have now taken over most of the phasing work.

So let me turn now to where 3D imaging stands today. In our cell work, a start has been made by David Shapiro and Pierre, with David obtaining not just the one stationary pattern that I showed at the outset, but a set of eight patterns 1° apart in their orientation, and Pierre then going on to phase those patterns and obtain the eight successive cellular views. He has put those into a little movie, which is available at the web site of PNAS at http://www.pnas.org/content/vol0/issue2005/images/data/0503305102/DC1/03305Movie_2.mov.

Note that in this work only eight planes in diffraction space have been obtained and phased. Thus, the step to a genuine 3D phased data set has yet to be taken by us. It has been taken on a 3D man-made specimen, somewhat akin to our two-dimensional (2D) specimen of 5 years ago, by Henry Chapman and his colleagues, also available on the web at http://www.pnas.org/content/vol0/issue2005/images/data/0503305102/DC1/03305Movie_1.mov. This movie is in a sense a little disappointing, in that with phased 3D data a crystallographer would probably have moved inside the specimen and shown much more than just another rotating image of the specimen from outside. The necessary data set is there however.

So, we see that progress on the two things that a person in 2000 would have wanted to see by 2005 – an advance to 3D and real biological specimens – is taking place. In an interesting development, we in fact expect to image the yeast cell in two different ways. Let me explain.

We are now planning an upgrade to our beamline at Berkeley which will allow it to give photons at wavelengths near 8 Å, as well as in its present softer range covering the "water window" extending from about 23 to 44 Å. The 8 Å photons have sufficient penetrating power that the Born approximation will be met by our 3 µm yeast cells, and normal 3D phasing and imaging can take place through the fine-sampling technique. This approach has the advantage that as cell biologists begin to want single-particle imaging, and want to graduate from small yeast cells to larger (e.g., human) cells, it will only be necessary to shift to more penetrating (such as 5- or 2.5 Å-wavelength) X-rays. Such a start in shorter wavelength diffraction, using *Escherichia coli*, has also recently been made at SPring8 in Japan by John Miao and his associates. With the 23 Å photons the penetrating power is less, and the Born condition does not accurately hold, but another condition, the Rytov approximation, which treats diffraction basically as a transmission rather than a scattering phenomenon, does still hold, and this leads to a different method of image formation, in which the phasing and imaging is carried out two-dimensionally, at each separate orientational setting of the specimen – just as in the eight-setting movie – and then assembled together in 3D space to obtain the desired 3D image; the process is, in short, a tomographic, rather than a conventional, 3D imaging method. Erice happened to come along at a time when we still have not produced the large number of 2D images needed for the 3D assembly process, but we believe that it will succeed when we do produce those images. This tomographic approach has been introduced into our thinking by Elser and Thibault at Cornell. The advantage of the technique, biologically speaking, is that the vitreous ice in a flash-frozen cell, while still present to preserve the original structure and protect it from radiation damage, is highly transparent to the water-window photons, and effectively disappears in the imaging, allowing the organic material to be displayed with increased clarity. Thus, we think that water-window imaging may become a sort of specialty method for the smaller cell types and for smaller cellular components, providing the highest quality of imaging, and as such very worth developing along with the more regular 3D type of imaging. In another year or so, we should have 3D images of the yeast cell by both of these imaging methods. When that occurs, biological science, we feel, will once more have received from crystallography a decisive new tool.

Acknowledgements

We acknowledge support provided by Bruce Futcher of Stony Brook University's Department of Biochemistry and Cell Biology for providing the cell line imaged with our microscope, by Lawrence Berkeley's Advanced

Light Source staff, and by Pierre Thibault for his help in preparing and presenting the figures for this talk. This project is supported by National Institutes of Health, National Science Foundation, and Department of Energy grants to the Stony Brook and Cornell groups. The Advanced Light Source at Lawrence Berkeley National Laboratory is supported by the Director, Office of Science, Office of Basic Energy Sciences, Scientific Facilities Division of the US Department of Energy.

References

1. Miao, J., Charalambous, C., Kirz, J., and Sayre, D. (1999) Extending the methodology of X-ray crystallography to allow imaging of micrometre-sized non-crystalline specimens. *Nature*, **400**: 342–344.
2. Shapiro, D., Thibault, P., Beetz, T., Elser, V., Howells, M., Jacobsen, C., Kirz, J., Lima, E., Miao, H., Neiman, A.M., and Sayre, D. (2005) Biological imaging by soft x-ray diffraction microscopy. *Proceedings of the National Academy of Sciences of the USA*, **102**: 15343–15346.
3. Sayre, D., Chapman, H.N., and Miao, J. (1998) On the extendibility of X-ray crystallography to noncrystals, *Acta Crystallographica*, **A54**: 232–239.
4. Sayre, D. (1980) Prospects for long-wavelength x-ray microscopy and diffraction. In *Imaging Processes and Coherence in Physics*, vol. 112. Edited by Schlenker, M. et al. Berlin: Springer, pp. 229–235.
5. Sayre, D. (1952) Some implications of a theorem due to Shannon, *Acta Crystallographica*, **5**: 843.
6. Sayre, D. (1991) Note on "superlarge" structures and their phase problem. In *Direct Methods of Solving Crystal Structures*, vol. 274. Edited by Schenk, H. New York: Plenum Press, pp. 353–356.

Printed in the United States
111970LV00001B/178-189/P